W9-CXM-910

Cultures in Orbit

CONSOLE-ING PASSIONS

TELEVISION AND CULTURAL POWER

Edited by Lynn Spigel

Lisa Parks

Cultures in Orbit

Satellites and the Televisual

DUKE UNIVERSITY PRESS   DURHAM AND LONDON

2005

© 2005 Duke University Press

All rights reserved

Printed in the United States of America on acid-free paper ∞

Designed by Rebecca Giménez

Typeset in Scala by Tseng Information Systems

Library of Congress Cataloging-in-Publication Data

appear on the last printed page of this book.

In loving memory of my grandmother,

Marion Mann.

Contents

Acknowledgments

This project began nearly a decade ago in a graduate seminar, and it has, thankfully, mutated and transformed many times since then. It once took shape as my Ph.D. dissertation in the Department of Communication Arts at the University of Wisconsin, Madison, where I was fortunate to work with a wonderful medley of professors and graduate students, whom I will never be able to thank enough for creating such a stimulating place to think, learn, and live. Faculty members Julie D'Acci, Michele Hilmes, Nick Mirzoeff, Jo Ellen Fair, David Bordwell, Vance Kepley, Robert McChesney, and Jack Kugelmass offered wisdom and encouragement in and out of their courses, and in one way or another they helped spawn ideas related to this project. I am especially thankful to my adviser and friend John Fiske, whose juggling of intellectual pursuits, political commitments, and popular pleasures dazzle me to this day and who helped spin me into the orbit of writing this book. Also instrumental was a sharp and lively cohort of fellow graduate students and friends at UW, Madison, including Doug Battema, Aniko Bodroghkozy, Carolyn Bronstein, Paula Chakravarrty, Steve Classen, Norma Coates, Jennifer Fay, Shari Goldin, Derek Kompare, Elana Levine, Daniel Marcus, Donald Mekiffe, Jason Mittell, Matthew Murray, Darrell Newton, Jeff Sconce, Chris Smith, Tasha Oren, and Pamela Wilson. Most of all I thank Moya Luckett, Jenny Thomas, Kevin Glynn, Andrew Foster, and Guven Sargin for remaining dear pals over the years. I am also extremely grateful to Michael Kackman, whose insights and warmth are scattered throughout these pages.

When I moved to the Department of Film Studies at the University of California, Santa Barbara, this book continued to develop and transform. My colleagues and students here have been tremendously generous, and I am privileged to work with people of rare intelligence, kindness, and collegiality. I am grateful to the Film Studies faculty: Edward Branigan, Dana

Driskel, Anna Everett, Dick Hebdige, Constance Penley, Bhaskar Sarkar, Janet Walker, Cristina Venegas, and Charles Wolfe. Staff members Kathy Carnahan, Joe Palladino, and Roman Baratiak enrich our workplace in innumerable ways with their knowledge, wit, and capacity for smooth operations. I have also had the fortune to work in interdisciplinary contexts with other UCSB faculty, including Bruce Bimber, Lisa Jevbratt, George Le Grady, Alan Liu, Marko Peljhan, Rita Raley, Celine Shimizu, Bill Warner, and Juliet Williams. I also thank my dedicated and hardworking UCSB research assistants Melissa McCartney, Lisa Fotheringham, Kim Lewis, and Elisa Iovine, as well as Julia Himberg for her friendship and scholarly enthusiasm.

In the process of writing this book I was fortunate to travel to several different places to conduct different phases of research. Many people helped direct me to relevant resources and materials. I thank the staff at the National Public Broadcasting Archives in College Park, Maryland, and the Library of Congress Motion Picture Archives in Washington, D.C., where I viewed *Our World*, which had to endure a transfer process from two-inch video to VHS in the archive's restoration department. I especially thank archivist Rosemary Haines, who kept me apprised of the transfer progress and helped arrange my visit. Julia Paech, Coralie Ferguson, Jose Anunciada, and Jacqueline Bethel facilitated my research activities at Imparja TV and CAAMA in Alice Springs, Australia, during the summer of 1999. I thank Imparja's staff members for allowing me to visit their facilities and observe their production processes and for taking the time to talk with me. My trip to Bosnia in 2001 was made possible by the generosity of Toma Longinovič, Aleksandar Bogdanič, and Beti Tomsič, as well as numerous Bosnian, Croatian, and Slovenian students and citizens and UN representatives whom I had the pleasure to meet and talk with along the way. I am also grateful to the staff at UCLA Medical Center, who brought me back to health when I returned from this trip ill, and to Anil de Mello for being an amazing friend during this difficult time.

Over the past several years I presented portions of this research in different settings, and there are many people I would like to thank for inspiring me with their work and/or influencing the project by reading or discussing portions of the manuscript with me. They include Shanti Kumar, Wolfgang Ernst, Ursula Biemann, Rosi Braidotti, Lynn Spigel, Jody Berland,

Caren Kaplan, Mischa Peters, Nina Lykke, Mette Bryld, John Hartley, Jon
Beller, Margaret Morse, Jim Schwoch, Mimi White, Anna McCarthy, Sherri
Rabinowitz, Kit Galloway, Brian Springer, Irit Rogoff, Ella Shohat, Berteke
Waaldij, Anneke Smelik, and Douglas Davis. I am also thankful to different institutions for inviting me to give talks, including the University of
California, Santa Cruz; University of Utrecht; University of Southern Denmark; University of Banja Luka; University of Ljubljana; the Suchbilder
Group in Berlin; the ITH in Zurich; the Basel School of Media Arts; the University of Wisconsin, Milwaukee; the University of Wisconsin, Madison;
and the Art Center in Pasadena. At all of these institutions I gained important insights and feedback. I am also indebted to the exuberant graduate
students in my Television Theory seminar at the University of Southern
California, where I taught as a visiting professor for a semester.

At Duke University Press I thank Ken Wissoker for being such an eru-
dite editor and supporter of this work and Christine Dahlin, Courtney
Berger, Mark Mastromarino, and Joe Abbott for editorial support during
the project's final stages. I am also deeply indebted to the anonymous
reviewers of the manuscript, who provided invaluable comments during
the revision process. The work also benefited from the support of several
grants, including a NASA/Wisconsin Space Grant Consortium Fellowship,
a UC Regents' Junior Faculty Fellowship, and a faculty research grant from
UCSB's Interdisciplinary Humanities Center.

Finally, I thank my family members scattered across the Montana moun-
tains, California suburbs, and New York metropolis, as well as my dear
friends from Missoula, Montana, Chip Stearns and Michel Valentin, who
pushed me to go into academia in the first place. Without their personal
libraries, passionate encouragement, and late-night conversations I would
never have been motivated to write such a book. Last but not least, I thank
Miha Vipotnik, whom I encountered in the strangest orbit but who has
become my most familiar, most intimate path.

An earlier version of chapter 1 appeared in *Planet TV: A Global Television
Reader*, Lisa Parks and Shanti Kumar, eds. (New York: New York University
Press, 2002), 74–93, and an earlier version of chapter 3 appeared in *Social
Identities*, 7:4, 2001, 585–611.

Cultures in Orbit

"M11" digital composite by Miha Vipotnik.
Courtesy of Miha Vipotnik.

Introduction

The first time I saw a satellite I was in Akumal, Mexico. It was a tiny moving blotch of light, more steadfast than a shooting star, more distant than an airplane. As I tracked the satellite's trajectory through the night sky, little did I know it would leave so many traces, triggering a web of curiosities about the barely visible machines encircling our planet. It was only after I caught sight of a satellite with my own eyes that I began to wonder, How many satellites are there? What do they do? Who controls them? And what can they see? At the time, I imagined the satellite to be as removed from everyday life as it appeared. But as I looked into the matter further, I found that of all communications technologies, satellites, perhaps paradoxically, have the tightest grip on our world. Over time, my leisurely glance into space became the labor and love of researching and writing this book. Throughout *Cultures in Orbit* I use the satellite's remote position on the fringes of what we see and know to extend, alter, and complicate the way we understand something more familiar—television. This book, in other words, is not a history of satellites. It does not discuss the technical properties of satellites in detail. It does not review the state or institutional regulation of satellites. Nor is it a survey of satellite television services around the world. Instead, it is an attempt to develop a critical strategy for understanding satellite television convergence and how particular satellite uses have reconfigured the meanings, practices, and potentials of the televisual.

Throughout the book I advance two main arguments. First, I use the phrase *satellite television* to refer not only to direct satellite broadcasting but to the convergent practices of live international transmission, remote sensing, and astronomical observation as well. I argue we should no longer think of television only as a system of global commercial entertainment or national public broadcasting because historically the medium has also been organized and used as a technology of military monitoring, public education, and scientific observation. Rather than limit the definition of the televisual to the technical apparatus or popular pleasures of broadcasting, then, I define it as an epistemological system derived through the alternating discursive modalities of commercial entertainment, public education, military monitoring, and scientific observation.[1]

Second, the satellite television practices discussed throughout the book—live international transmission, direct satellite broadcasting, remote sensing, and astronomical observation—have helped to determine (that is, to shape and set the limits of) the spheres of cultural and economic activity that constitute what we know as "the global." Without satellites it would be impossible to photograph the world as a "whole earth," to watch the same television program at the same time on different continents, to see the earth in a cosmic context. As Fredric Jameson suggests, "the intellectual space of globalization" is "a space of tension, in which the very 'problematic' of globalization still remains to be produced."[2] In an effort to produce the global as a space of tension, this book explores, on the one hand, how satellite television has been used to construct and affirm "global village" discourses that envision the world as a unified, organically evolving cultural, economic, political system, perhaps symbolized best in "Earth-shot" images that work to synthesize, contain, and transform the world's irreducibility into an iconic expression of global totality. On the other hand, the book considers how those same satellite television practices have cut and divided the planet in ways that support the cultural and economic hegemony of the (post)industrial West, whether through the inscription of satellite footprints, uneven coverage of world events, or the maintenance of particular epistemological boundaries.

Overall, then, the book can be understood as an attempt to complicate and rethink the meanings of and relations between "television" and "the global" through analyses of different configurations and uses of satellite television. Before presenting these analyses, I would like to briefly delineate the satellite industries that have emerged since the launch of *Sputnik* in 1957 and map the critical constellations that inform this project.

Satellite Crossings

When *Sputnik* was launched in 1957, people around the world turned toward the skies to witness Russia's historic satellite circle the globe. American satellites *Echo 1, Telstar,* and *Early Bird* followed in the early 1960s.

These launchings made possible a series of live satellite television experiments in the 1960s including the Tokyo Olympics (1964), a Sotheby's art auction (1965), and *The Town Meeting of the World* (1965). By the 1970s, live satellite feeds became standard in the television industry as reporters covered events from the Jonestown massacre in Uganda to the war in Vietnam to airplane hijackings in the Middle East. During the 1980s the uses of satellites proliferated as multinational media conglomerates, nation-states, and nonprofit organizations alike began using them to relay television signals across the planet. At the beginning of the twenty-first century there were an estimated eight thousand satellites in orbit. Satellites are now at the core of our global telecommunications infrastructure, and they have become a principal means by which we see and know the world and the cosmos beyond.[3]

Satellite dishes speckle rooftops and apartment buildings in many parts of the world, suggesting there are more direct satellite broadcasting services and viewers than ever before. Satellite television networks such as Rupert Murdoch's Star TV and British Sky Broadcasting have attracted audiences in Asia and Europe respectively. Univision and Telemundo serve Spanish-speaking viewers throughout the Western Hemisphere. In the United States Primestar, DirecTV, Dish Network, and U.S. Satellite Broadcasting have jockeyed for control over North American markets. Global television news networks such as CNN and BBC World would be impossible without satellites. And a number of national direct satellite broadcasting services have emerged to compete with the lords of the global village, including British Satellite Broadcasting, India's Doordarshan International, and Chicago Chinese Communications, Inc.[4] Al-Jazeera beams its signal, the only "uncensored information and commentary from an Arab perspective," to thirty million satellite television viewers worldwide.[5] In Senegal the French media company Canal Horizons delivers foreign entertainment while promising to support African coproductions.[6] Indigenous groups in Canada and Australia have developed their own satellite television services.[7] In short, the globe is crisscrossed by satellite footprints, and the meanings of the televisual are increasingly contingent on them.

Remote sensing is another satellite industry that underwent rapid commercial expansion during the 1990s. In 1994 U.S. president Bill Clinton implemented a remote-sensing policy that for the first time allowed private U.S. companies to sell high-resolution satellite imagery to customers at home and abroad.[8] For decades only military intelligence officers and scientists had access to such satellite images, some of which were generated by *Satellite Pour l'Observation de la Terra* (SPOT), a French satellite that has led the way in commercial remote sensing since the mid-1980s. During the past decade, U.S. companies have formed multimillion-dollar partnerships to take advantage of this expanding field in remote sensing, a field that involves the convergence of satellite, television, and computer industries. Space Imaging, Inc., for instance, is a $500 million joint venture of E-Systems, Mitsubishi, Kodak, and Lockheed Martin. And a smaller company called Earthwatch, Inc., has partnered with Hitachi, 3M Images, Ball Aerospace, and Worldview Imaging to capitalize on the growing demand for remote-sensing data. Television networks have broadcast meteorological satellite images since 1960, but high-resolution satellite views of the earth are now appearing in television news as well.[9] The first network to televise such an image was ABC when, in 1989, it purchased and aired a SPOT photo to provide coverage of a Libyan chemical weapons facility.[10] Executives at ABC boast that they can use satellite images to "fly" viewers directly through battle zones during evening news reports. Similar images have also been used by U.S. officials to count the number of African American men who participated in the Million Man March in Washington, D.C., in October 1996. In addition, satellite images were used by prosecutors in the O. J. Simpson trial in their harried attempts to locate Simpson's white Ford Bronco on the day of Nicole Simpson's murder. Remote sensing has even been used by archaeologists to excavate ancient sites around the world, from Mayan ruins in Mexico to Cleopatra's palace in Egypt. Finally, satellite technology has provided military commanders and intelligence officers with strategic platforms from which to monitor theaters of war and civil strife. Remote sensing is related to the televisual, then, for it involves practices of seeing and knowing across vast distances and can powerfully shape our worldviews and knowledges of global conflicts, histories, and environments.

Just as satellites have helped to shape our worldviews, they have mediated our understanding of the earth's relation to its celestial surroundings. Over the course of the past forty years NASA, the European Space Agency, and the USSR deployed a host of astronomical satellites to study outer space

phenomena. Satellites such as *Voyager, Magellan, Ariel, Hipparcos,* the MIR
space station, and the Hubble Space Telescope have been used to produce
astronomical observations, whether orbiting our planet or others. Like remote sensing, astronomical satellites gather image data about distant matter and facilitate vision through time and space. But since they are turned
toward outer space rather than the earth's surface, they represent phenomena billions of light years away. Astronomical satellites thus extend the
televisual by making matter at an almost incomprehensible distance and
scale intelligible, extending the domain through which distant vision and
knowledge are possible.

Astronomical satellite images have also appeared in television broadcasts as astronomers and reporters excitedly illustrate new space findings
and attempt to place the earth in a cosmic context. The public display of
astronomical images is important since these images come from the few
satellites that remain in the public domain, funded with taxpayer dollars.
But Dennis Tito's $20 million space tour aboard the International Space
Station and the efforts of Space Marketing, Inc., to deploy billboards in low
earth orbits suggest we may be well on our way toward further privatiza-
tion of orbital and outer space.[11] Satellites, then, are not just extensions
of scientific institutions but have been imagined as technologies of space
tourism and advertising as well.

The Satellite in Cultural Theory

Despite the global significance of satellite technologies, cultural theorists
have been relatively silent about their ramifications. Prominent contemporary thinkers have alluded to the satellite's relevance to globalization, but
few have developed a sustained discussion or critique of the technology's
uses or effects. In his effort to identify possible origins of the current phase
of globalization, Anthony Giddens suggests that

> if one wanted to fix its specific point of origin, it would be the first
> successful broadcast transmission made via satellite. From this time
> onwards, instantaneous electronic communication across the globe is
> not only possible, but almost immediately begins to enter the lives of
> many millions. Not only can everyone now see the same images at the

same time; instantaneous global communication penetrates the tissue of everyday experience and starts to restructure it—although becoming restructured in turn; as a continuous process.[12]

While Giddens recognizes the satellite's unique relation to globalizing processes—specifically, its structuration of everyday experience—he implicitly suggests that people around the globe experienced this moment in the same way. Further, we are left with no knowledge of what was actually broadcast or how the capacity for instant global connection restructures the "tissue of everyday experience" to which he refers.

In *Simulations* Jean Baudrillard makes oblique references to satellite technologies, using terms such as *orbital* and *satellisation* as metaphors for the highly controlled social order that emerges in the age of nuclear proliferation. For him the satellite is a symptom of the West's quest for strategic planetary control: He offers the term *satellisation*, to refer to the way systems of technological security and control (that emerged with nuclear deterrence) determine and are reflected within the social order. He writes, "This same model of planned infallibility, of maximal security and deterrence (which satellites were designed to produce and enforce), now governs the spread of the social. That is the true nuclear fallout: the meticulous operation of technology serves as a model for the meticulous operation of the social."[13] Satellites suggest a "universe purged of every threat to the senses, in a state of asepsis and weightlessness—it is this very perfection which is fascinating."[14] Baudrillard suggests that the satellite's emergence is symptomatic of a military paradigm that overdetermines modes of social organization. While he recognizes the satellite's relation to material conditions on the earth, he does so ultimately to prove a broader point about the social effects of nuclear proliferation.

Where Giddens and Baudrillard explicitly refer to the satellite, even if in passing and metaphorical ways, some theorists of globalization sidestep it altogether. In his important and widely read book *Modernity at Large: Cultural Dimensions of Globalization*, Arjun Appadurai develops a framework for analyzing global cultural flows without ever mentioning satellite technology. How is it, we might ask, that the global flows, which Appadurai identifies as ethnoscapes, mediascapes, technoscapes, financialscapes, and ideoscapes, are able even to be imagined much less actualized without the

existence of the satellite? One need only glance through recent collections
such as Jameson's and Miyoshi's *The Cultures of Globalization* or Wilson's
and Dissanayake's *Global/Local*, which feature other heavy-hitter cultural
theorists, to find that despite their very insightful discussions of globalization as a historical, cultural, and economic phenomenon, the satellite does
not even make it into the index.

In cultural theory the satellite has been missing in action, lying at the
threshold of everyday visibility and critical attention, but moving persis-
tently through orbit, structuring the global imaginary, the socioeconomic
order, and the tissue of everyday experience across the planet. This blind-
ness is not innocent, for it reveals that the military-industrial-information
complexes of the West have been quite effective at concealing and using
their most strategic technologies to assume global domination in the post–
cold war period, managing to avert the critical gaze in the process. This
is especially troubling since the satellite represents such a colossal accumulation of the very forms of industrial, military, and scientific capital
and power and knowledge that have so preoccupied leftist intellectuals and
sociocultural critics of globalization. The satellite is arguably one of the
most elaborated forms of capitalism since it embodies immense financial
investments, years of strategic planning, and pools of scientific labor. Despite its lofty requirements the satellite has floated above the critical horizon, functioning as a structuring absence in cultural theory. Its uses structure and reflect global material conditions that fixate us, but these uses have
themselves remained in their own orbit.

The Satellite in Television and Cultural Studies

Cultures in Orbit embraces the satellite's status as a structuring absence, a
technology on the perimeter of everyday visibilities and cultural theory, as
a way to engage and extend critical discussions of the televisual. For even
in Television and Cultural Studies, where the satellite might seem a bit
closer to home, there have been only a handful of studies. Scholars in fields
such as international and mass communications and political science have
been discussing the institutions and regulation of satellite communication
since the 1960s, but very few have analyzed the technology from a critical
perspective. Most such research has emerged from scholars in England,

Canada, and East Asia, a result, in part, of the proliferation of direct satellite broadcasting services in these regions during the 1980s and 1990s and
ensuing concerns about the "Americanization" of national television cultures. This work has established preliminary frameworks for further analysis of satellite technologies, combining media sociology, national cultural
politics, and science and technology studies.

Media sociologist Shaun Moores used ethnographic methods to study
the reception of direct satellite broadcasting services in different homes
and neighborhoods in England. Moores's research was motivated in part
by his assessment that "what remains totally absent from this [scholarly]
literature [on satellite broadcasting] is any understanding of the significance that satellite TV has for consumers in everyday contexts."[15] He was
especially interested in the kinds of mobility satellite TV offers to viewers, asking, "To what new destinations is it promising transport and who
chooses to make the journey?"[16] In a related study, "Satellite Dishes and
the Landscapes of Taste," Charlotte Brunsdon analyzes the installation of
satellite television dishes on English homes during the late 1980s and early
1990s in relation to social struggles over public space, cultural taste, and
national broadcasting policy.[17] In the context of British Sky Broadcasting's
expansion, the satellite dish became a culturally loaded object; its location
in public view on home exteriors and its association with working-class
taste formations triggered conflicts over the class-based elitism of British
broadcasting policy. Finally, in a creative departure from the study of satellite broadcasting, Canadian scholar Jody Berland analyzed scientific satellite practices such as remote sensing within a cultural studies paradigm.
She conceives remote-sensing satellites not only as technologies of meteorology and espionage but also as deeply embroiled in changing visuality,
national culture, and scientific prerogatives. Often represented as a "great
white north" on the edge of the meteorological frame or as a dense mass of
white clouds, she argues, satellite images position Canada as a nonplace —
a barren expanse that exists only as a symbolic exterior to the heartland of
the United States.[18] Berland's work is important because it explains how
we come to understand images that arise from "the complex imperatives
and alliances of three interdependent industries: paramilitary space exploration; computer software; and television."[19]

Cultures in Orbit attempts to build on Television and Cultural Studies research on satellite television, and it is especially indebted to Berland's bold endeavor to place the topic of remote sensing within this critical rubric. Just as cultural theorists have considered satellites only in passing or metaphorically, until recently most Television and Cultural Studies scholars have overlooked their military and scientific uses, focusing primarily on direct satellite broadcasting and global entertainment. Berland has been the only media scholar to articulate scientific and military uses of remote-sensing satellites as part of changing televisual cultures and national identities. In one sense her work synthesizes the critical insights of Baudrillard, who connects military satellites to the emergence of a nuclear social order, and Brunsdon, who maps national cultural politics across the landscape of the satellite dish. *Cultures in Orbit* sets out to forge a similar constellation of unlikely bedfellows by analyzing military, scientific, and cultural uses of satellites alongside one another with a specific goal in mind: to rethink, complicate, and extend critical definitions of the televisual.

It is especially when technologies converge that we notice and understand their definitions. For in these moments of convergence we must be able to differentiate technologies in order to understand how they are coming together. Convergence involves not only the collision of industries and technical recombinations; it also involves shifts in the discursive construction of technologies that preexist the convergence and those that emerge as a result of it. In other words, convergence can be understood as an ongoing process rather than a radical break, sudden transformation, or turning point in technological formations.

In his famous book *Television: Technology and Cultural Form*, Raymond Williams offers a flexible and historically contingent definition of television, suggesting it is a cultural technology whose form is determined through its content and use. As he reminds us, television emerged as "a combination and development of earlier forms: the newspaper, the public meeting, the educational class, the theater, the cinema, etc."[20] As a derivative technology, television has always been a site of convergence.[21] If television emerges as a medley of earlier cultural forms, then we need to consider how its content and form changes in the age of the satellite. *Tele-*

vision was published in 1974, less than a decade after the first satellite television experiments marked the convergence of the orbital and homebound
machines. Williams had the foresight to predict that satellites "could have
the greatest impact on existing [television] institutions" because they will
likely "be used to penetrate or circumvent existing national broadcasting
systems, in the name of 'internationalism' but in reality in the service of
one or two dominant cultures."[22] Although skeptical about the possibility
of a genuinely "internationalist" television, Williams simultaneously proclaimed that "a world-wide television service, with genuinely open skies,
would be an enormous gain to the peoples of the world."[23]

Williams recognized the satellite's fundamental contradiction—that because of its technical complexity and high capital expense it would likely require the support of already dominant institutions, but it was nevertheless
a technology that could offer "enormous gain" to the world's people. Such a
proposition is consistent with Williams's more generalized critique of technological determinism and especially of his insistence that alternative de-
ployments of emerging technological systems might sustain and develop
subordinated cultures.[24] In a sense this project begins with the contradiction to which Williams alludes. That is, how might Western-controlled
satellite technologies be appropriated and used in the interests of a wider
range of social formations?

This question became prescient during the 1990s in relation to the Internet as progressives suggested the technology would democratize communication by virtue of its decentralized global network infrastructure
and hypertext features. Many digital enthusiasts, however, tended to leap-
frog the histories of preexisting networked technologies such as telephony,
radio, television, and satellites, assuming that since their institutional control landed in corporate hands, their uses were already fixed. If social constructivist approaches to the study of technology have taught us anything,
it is that technologies are dynamic, historically and socially specific systems involving a variety of actors.[25] We still have much to learn about the
wide range of actors that have given shape to these technologies, but this is
especially true of the satellite, since it is rarely ever inserted within the purview of the social, much less tied to subordinated formations. Further, as
Andrew Feenberg points out, in constructivist research "social resistance
is rarely discussed, with the result that [such] research is often skewed

toward a few official actors whose interventions are easy to document."[26]

Rather than focus on "official actors," *Cultures in Orbit* contributes to the social construction of the satellite by considering how a range of actors, including broadcasters, indigenous communities, military strategists, archaeologists, and astronomers, have used the technology in different ways in different milieus and, in the process, have altered the content, form, and meanings of television.

Raymond Williams's discussion of television is relevant for two other reasons as well. First, I adapt his critical approach to television as a cultural technology to study the satellite, considering how its form has been determined through content and uses. Just as television has historically been misunderstood as the act of transmission, so, too, have satellites been described as mere relay towers in outer space.[27] As I demonstrate throughout this book, however, satellites are involved in several televisual practices, including *live international television production, direct satellite broadcasting, remote sensing,* and *astronomical observation.* As technologies of cultural production, satellites generate and circulate televisual discourses. Second, these satellite practices have further elaborated what Williams understood as one of television's "innovating forms"—visual mobility—by moving its gaze across, beneath, and beyond the earth in unprecedented ways. Williams perceived visual mobility "as one of the primary processes of the technology itself, and one that may come to have increasing importance."[28] Throughout the project I consider visual mobility as a form of power and knowledge that is constituted by and through sites of satellite and television convergence. In other words, I articulate this potential for visual mobility with broader epistemological structures, especially those of Western cultural, corporate, military, and scientific institutions.

Because I am interested in how the meanings of satellite technologies take shape in the cultural forms and content they generate, each chapter offers analyses of media texts and discourses that inform and surround them. By examining *satellite television* as a site of convergence, I hope to complicate and refigure the definitions of each technology, exploring how their meanings have formed and shifted in relation to one another. To do so, I employ discourse and textual analysis to critically examine various kinds of "satellite content," including television shows prepared for live international broadcast, programming or flow packaged for direct satel-

lite broadcast, remote-sensing images, global-positioning maps, and astronomical images. By engaging with these orbital cultures—analogue and
digital sounds and images relayed and generated via satellite—I hope to
establish the satellite's relationship to the changing meanings of television.

Especially important to such an endeavor are the critical nuances of the
terms *television, tele-vision, televisuality,* and *the televisual,* which I want to
briefly delineate since I will be drawing on and working through them. I
use the term *television* broadly to refer to the established commercial and
public broadcasting institutions and the production and reception of programming that define them. In some cases I use *tele-vision* to refer to a
practice of remote seeing, which may not emanate directly from these institutions. For instance, I suggest that technologies including the telescope,
the computer, and the satellite enable or simulate distant sight and thus
participate in tele-vision. The term *televisuality* can be defined as the complex meanings and ideologies articulated by the images, programming,
flow, or coverage that is arranged and packaged for television transmission,
whether distributed over the air, via satellite, or online. As John Caldwell
suggests, televisuality also involves (among other things) the stylistic excesses and self-exhibitionism of the medium—that is, a particular presentational manner that is contingent on sociohistorical, political, economic,
and technological conditions.[29] Finally, when I use the phrase *the televisual* I
am referring to different structures of the imaginary and/or epistemological structures that have radiated from and taken shape around the medium
over its history. The televisual is a particular set of knowledge practices or
ways of seeing and knowing the world that are not necessarily bound to
television industries. The televisual, in other words, is an epistemological
system that can be activated within and across different discursive fields
such as archaeology, geography, or astronomy. The televisual can finally be
considered as a set of critical discourses that define and attribute properties
to the medium—for instance, as one of liveness, presence, flow, coverage,
or remote control.

While this book is informed by the critical paradigm of Television and
Cultural Studies, it draws on and combines ideas across various disciplines
including feminist criticism, science and technology studies, and cultural
geography. It employs a model of cultural studies understood "as both a
practice and a promise of interdisciplinarity attentive to epistemologies,

asymmetries of power, and forms of embodiment at issue in knowledge production, be it in the life and physical sciences, social sciences, humanities or outside academic and research centers."[30] It tries to formulate a way of understanding television that takes into account its historical and ongoing convergences, its migration across different disciplines and geographic places, and its relationship to military strategy and scientific knowledge. One of the goals of the project is to use satellites to move the study of television out of its proper place, beyond the nation, the broadcast institution, and the home.[31]

The interdisciplinary scope of the project has generated a need to imagine television not only as a site of commercial entertainment but as a site of military intelligence and scientific observation as well. Decades of satellite uses have shaped not only *what we see on television* but also *how we understand what "television" is and means.* As Raymond Williams suggests, "New technology is itself a product of a particular social system, and will be developed as an apparently autonomous process of innovation only to the extent that we fail to identify and challenge its real agencies."[32] This observation is especially important in relation to satellites because as quintessential technologies of the cold war their "real agencies" have historically been shaped by military, scientific, and corporate interests, with the public interest being largely left by the wayside.[33] Because of this, the satellite has occupied a remote and obscure place not only in cultural theory but within the public or popular imaginary as well, seemingly beyond reach because it is so capital intensive, so tagged to national security, so under the control of scientific experts. If anything, this book is an attempt to wrestle the satellite out of the orbit of its "real agencies" so that it can be opened to a wider range of social, cultural, artistic, and activist practices. As we will see, in some instances this process involves appropriating television's currency as a popular form to challenge the authority of military, corporate, and scientific institutions over the satellite, or imagining television in different places to expose how it has either been appropriated by or has itself informed military and scientific epistemologies.

In this sense *Cultures in Orbit* is inspired by feminist critiques of technoscience as well, which set out to deconstruct the ways scientific knowledge and technologies are used to sustain and reinforce Enlightenment discourses, global militarization, and gendered power hierarchies. When I

invoke the words *science* or *scientific knowledge* throughout the book, I am
not referring to what Katherine Hayles calls the "sciences of complexity."[34]
Instead, I am making specific reference to a discourse of scientific rationalism that is associated with the intellectual project of the Enlightenment,
whose authority has historically been predicated on vision or sight. This
discourse posits science as a practice of detached observation in which one
identifies objects of knowledge from a distance. Feminists have critiqued
scientific rational discourse for reinforcing gendered oppositions between
the knower and known, subject and object, and mind and body (among
other things).[35] As Rosi Braidotti observes, such a "position produces the
idea of neutrality and objectivity in the sense of allowing for no particularity
about the site of observation."[36]

Cultures in Orbit is particularly concerned with the way that different
forms of *satellite television* have been used by states, scientists, and broadcasters to disembody vision and construct seemingly omniscient and objective structures of seeing and knowing the world, or worldviews. Such
practices occur, for instance, within live global television programs, satellite images of war zones and ancient ruins, and cosmic zooms through
deep space, and they often support and sustain scientific rational paradigms by positing the world (or the cosmos) as the rightful domain of
Western vision, knowledge, and control. As Donna Haraway reminds us,
this "gaze from nowhere" is "tied to militarism, capitalism, colonialism,
and male supremacy—[and it seeks] to distance the knowing subject from
everybody and everything in the interests of unfettered power."[37] To com-
bat such logic, Haraway encourages feminists to adopt a contradictory
position of embodied objectivity "that privileges contestation, deconstruc-
tion, passionate construction, webbed connections and hope for transformation of systems of knowledge and ways of seeing."[38] In one sense this
book is an attempt to partake in such a practice—to ground, materialize,
and embody satellite-facilitated vision and knowledge and to insist on its
partiality.

To emphasize this partiality, I have organized the book as a genealogy
rather than a history. Michel Foucault describes the practice of genealogy
as an alternative to history, insisting it should energize past conditions by
privileging instability and discontinuity over the acquisition of total knowledge and comprehension.[39] Foucault further suggests that the genealogist

moves from a perspective of distance to closeness, exploring the surfaces
of the particular but always from an alienated view. The satellite's physical
distance from the earth, its outpost position within the humanities disciplines, and its technologized gaze make it a compelling (if unlikely) alienated position from which to write an "effective" history or genealogy. For the satellite forces us to contemplate what it would mean to produce an account of the earth as alienated from it. The satellite's uniqueness lies in its orbital position, but this detachment from the earth need not result in the kind of distant observation valorized by Enlightenment discourse. Instead, this orbital position can be imagined as one of alienation or difference, which may catalyze desires for proximity, intelligibility, and connection as opposed to remote control.

As the book explores specific satellite uses and actors, consistent with the social constructivist approach to the study of technology, its organization as a genealogy is meant to foster reflection on the disparate yet connected paths through which current technological and cultural conditions have emerged. The genealogy of satellite television might even be framed as an act of global positioning. But rather than use a GPS receiver to orient an individual in global time and space, genealogy offers a way of positioning readers in the unwieldy occurrences that make up relations between present and past. Genealogy opens the field of the technological to a broader set of historical and cultural questions that energize new imaginaries rather than restricting our understanding to technical inventions or institutional regulations.

To delineate the book's trajectories, I open each chapter with global positioning coordinates that correspond to the various sites of analysis. These visualizations are offered as symbolic reminders of the book's endeavor to situate satellite practices within material conditions on the earth as well as serve as a playful *detournement* of global positioning practices that were designed to enhance military surgical strikes, not orient readers through text. These visualized positions are also meant to challenge the satellite's "view from nowhere" by locating the reader in a mediated position, a place-based imaginary, a site of geographic knowledge, all of which are increasingly produced via satellite.

What follows is an assemblage of case studies designed to emphasize the wide range of contexts in which satellites have been used. Since I am

interested in the way satellites have been used both to reinforce and challenge (in limited ways) the hegemony of the West, I have selected sites of
analysis that allow me to highlight how they have triggered instabilities and
tensions while also serving as a global security blanket. Each site of analysis is offered as a particular arrangement of "satellite television" that has
also been informed and determined in one way or another by epistemological boundaries such as West/East, modern/primitive, science/culture,
north/south, global/local, earthly/otherworldly, or feminine/masculine. In
each chapter I consider how satellites have been mobilized to negotiate
these tensions, often (but not always) in ways that privilege the industrial
West or reinforce assumptions that have sedimented in the Western imaginary. I use the terms *the West* or *Western* to refer to specific Eurocentric
positions or discourses that have been either extended or exposed by satellite television practices.[40] Throughout the book I also engage critical discourses from Television Studies to explore how satellite uses reconfigured
the meanings of such televisual terms as *liveness, presence, flow, footprint,
views, coverage, the gaze,* and *remote control.* Each chapter (1) describes a particular instance of satellite use; (2) discusses the actors, institutions, or
epistemological practices involved in its use; (3) analyzes media texts that
are circulated, generated, or inspired by satellites; (4) considers the political implications of their production and circulation; and (5) offers an
elaboration of existing concepts in Television and Cultural Studies.

Chapter 1 takes the reader to the BBC headquarters in London during
the 1960s, where broadcasters pulled off the first live international satellite
television program, *Our World* (1967), which was broadcast to an estimated
five hundred million viewers in twenty-four different countries.[41] Organized around the theme of the population explosion, this satellite spectacular celebrated the primacy of the West by contrasting scenes of an industrial
and culturally prolific North to an "overpopulated," "underdeveloped," and
largely unrepresented South. *Our World's* content articulated the liveness
of satellite television with the modernity, permanence, and civilizational
processes of industrial nations, undermining the utopian assumption that
satellites inevitably turned the world into a harmonic "global village." The
chapter closes with a discussion of how *Our World* reconfigured "liveness,"
interpellating the viewer not only as "globally present" but as "culturally
worldly" and "geographically mobile" as well. The televisual structures em-

bedded within this early satellite spectacular still persist and prefigure the

rise of global TV networks such as CNN and BBC World.

Chapter 2 shifts the discussion away from the Western control of satellite television and examines the programming and distribution practices of Imparja TV, an Aboriginal satellite television network headquartered in Alice Springs, Australia. In the 1980s Australian Aboriginals struggled for control over a transponder on the national satellite in an effort to regulate which television signals would come into their territories. The case of Imparja TV pushes us to rethink the concepts of television flow and the satellite footprint and to conceptualize them in relation to postcolonial conditions, cultural hybridities, and global media economies. I also discuss how Imparja's footprint and flow are implicated within ongoing Aboriginal struggles for territorial reclamation and cultural survival in Australia.

While chapters 1 and 2 examine satellite uses that are conventionally associated with the term *satellite television*, chapters 3 and 4 explore an altogether different satellite practice—that of remote sensing. In chapter 3 I discuss the production and circulation of U.S. military satellite images of the war in Bosnia. Focusing on television news segments containing U.S. satellite images of mass graves in Srebrenica, I explain how state satellite intelligence became part of the commercial television news coverage of wartime atrocities. In an effort to complicate the "objective" status of these images, I explore what it might mean to witness events from the perspective of an orbiting satellite, and I suggest the need for greater public knowledge of satellite technologies and increased literacy around the images they generate. In an age of state satellite coverage of war, one of the most important functions of the witness is to demilitarize such perspectives—that is, to open the satellite image to a wider range of critical practices and uses.

Whereas chapter 3 considers how satellites were used to monitor a global "trouble spot," chapter 4 describes how they have been used to locate and excavate one of the so-called bedrocks of Western civilization, Cleopatra's palace in Alexandria, Egypt. This chapter explores how archaeologists use remote sensing to peer back in time and read the earth's surface as a text. It also explores how the televisual permeates scientific disciplines organized around practices of distant discovery. Rather than embrace the technologically determinist claim that satellites simply enhance and improve archaeologists' vision, I analyze the use of satellites in relation to various cultural discourses that construct Cleopatra as a sexual spectacle, site of racial ambiguity, and monument of Western civilization. The remote sensing of Cleopatra, then, is not just a technologized archaeological excavation; it involves the compulsion to imagine the ancient queen as a sexual spectacle that can be materialized and touched in the present using satellite and computer vision.

Finally, chapter 5 moves the satellite's gaze away from the surface of the earth and out into deep space, exploring how the Hubble Space Telescope's astronomical observations are implicated in televisual practices and Western fantasies of remote control. In this chapter I discuss a range of visualizing practices that Hubble images invoke and participate in, including the live media event, the sonogram, the digital effect, and the science fiction film. I examine Hubble images of the Shoemaker-Levy comet's collision with Jupiter, of nebulae or "stellar nurseries" in deep space, and their integration within documentary and science fiction films such as *Cosmic Voyage, Contact,* and *The Arrival.* I offer the term *satellite panorama* to refer not to a tranquil vista of outer space but rather to the contested discursive terrain that Hubble images activate as the conventions of astronomy, live satellite television, and digital imaging are brought to bear in their circulation and interpretation. The further Hubble gazes into space, the more tightly its otherworldly images are tethered to human embodiment and world history. This discursive strategy of remote control, I argue, is a key mechanism for regulating the unknowingness and uncertainty surrounding Hubble's digital and televisual views of outer space.

Gathering materials for this project took me into an almost global orbit, from the Hornbake Library's Public Broadcasting Archives at the University of Maryland to Imparja TV headquarters in Alice Springs, Australia, from the Library of Congress Motion Picture Archives in Washington to postwar villages of former Yugoslavia, from the NASA History Office to Egyptian archaeological excavations presented on the World Wide Web. In instances in which I write about a culture or place other than my own, I made efforts to visit and talk to people to form impressions that could be used to contradict and complicate the satellite's distant position and Western views. While the voices of Aboriginal broadcasters and Bosnian civilians may not be at the center of this text, they profoundly shaped and influenced my discussion of satellites and television. My visits to central

Australia and Bosnia reminded me that despite my critical preoccupation

with the technologization of vision, there is nothing like being with people
and seeing places and things with one's own eyes. This is, after all, where
Cultures in Orbit began—the moment I saw a satellite for the first time. Perhaps, then, this book is ultimately about a need or utopian desire to imagine
a technologized form of proximity, sensuality, and mutuality across borders that is not overshadowed by military strategies, scientific quests, and
corporate calculations. Unfortunately, the sociohistorical forces that constitute the West and its agendas make such a proposition extremely difficult
and unlikely and yet, perhaps, all the more important in an era of bounded
globalizations.

Satellite Spectacular

Designed by BBC artist German Fecetti, the logo for *Our World*—one of the first live international satellite television programs—incorporates a Da Vinci–inspired figure mapped over the earth's grids of longitude and latitude, its arms encircling the globe. In an evocative statement that collapsed global travel and world history within Da Vinci's iconic image of Western rational intellect, one of the show's producers declared, "It took three years of his life for Magellan to go around the world. The Graf Zeppelin took three weeks. A Russian cosmonaut made it in 90 minutes. . . . We are in a sense, electronic Magellans."[1] The "electronic Magellan" not only became a powerful metaphor for the way that satellite technology promised to ricochet *Our World*'s viewers around the globe from the comfort of the living room, but it also revealed how broadcasters in the industrial West imagined new technologies of space communication.

From 1962 and 1967 broadcasters in Western industrial nations participated in a series of live international television exchanges using the *Telstar, Early Bird*, and *Syncom* satellites. These live-via-satellite television programs, which I refer to as "satellite spectaculars," began just after the launch of the first United States commercial satellite, *Telstar*, in July 1962, and continued throughout the decade. While *Telstar* forged satellite connections across the Atlantic in a series of exchanges between the United States and Western Europe, *Syncom* established a satellite link across the Pacific, integrating East Asia within an expanding global satellite system. In 1964 Japan relayed the opening ceremony of the Olympics live to viewers across the Pacific. In 1965 a program called *The Town Meeting of the World* was shown live via *Early Bird* and beamed across the Atlantic. But the most ambitious satellite spectacular of the decade was the 1967 broadcast of *Our World.*

What distinguished *Our World* from earlier satellite broadcasts was its deliberately global reach: It was intended to link nations across the Pacific and the Atlantic, the communist East and the democratic West, the industrialized North and the developing South. In addition, *Our World*'s producers fully exploited what they understood to be the unique properties of live satellite television: its capacity to craft a "global now." Described by critics as a "fabulous planetary swing," a "spectacular display of electronic wizardry," "a vast global happening," and "an old fashioned geography class gone electric," *Our World* was relayed live via satellite on June 25, 1967, to an estimated five hundred million viewers in twenty-four countries.[2] The two-hour show, coordinated by the European Broadcasting Union and edited from master control at the BBC in London, required more than two years of planning, ten thousand technicians, four satellites, thousands of miles of land lines, and five million dollars to produce. The show predicated itself on the cultural legitimacy of public broadcasting, the benevolent paternalism of Western liberals, and the space-age utopianism of satellite communication. *Our World*'s structure established precedents for subsequent global television coverage, alternating between live views of the television studio, maps, and remote locations, interpellating the viewer not only as "globally present" but as "culturally worldly" and "geographically mobile." Seen by millions of viewers throughout North America and Europe, this early experiment helped to determine one of the cultural forms of satellite television.

This chapter examines the content of and discourses surrounding this early satellite spectacular in an effort to understand the particular forms of televisuality it generated. John Caldwell uses the term *televisuality* to refer to the aesthetic excesses that characterized U.S. television during the 1980s and 1990s, a time he describes as "an important historical moment in television's presentational manner, one defined by excessive stylization and visual exhibitionism."[3] Such stylistic excesses and visual exhibitionism can be recognized in earlier moments of television's history as well, particularly in moments of its convergence with other technologies. In the 1960s *Our World* and other satellite spectaculars constantly called attention to their own immediacy and liveness, aggressively using mise-en-scène, graphics, narration, and publicity to construct a form of television that was

imagined as different from earlier forms.

Generated at the peak of the cold war, in the midst of the space race, and during the decolonization of the developing world, satellite television first took shape in a series of broadcasts emanating from the United States, Western Europe, and Japan. These broadcasts exploited "liveness" as their defining feature and articulated it with Western discourses of moderniza- tion, global unity, and planetary control. By analyzing discourses surround- ing this experimental broadcast I hope to complicate the technologically determinist assumption that satellites simply extended television's global reach, further elaborated its capacity for "liveness," and created a harmo- nious global village. Not simply an aesthetic, satellite televisuality was also the result of complex and dispersed industrial practices, namely a decen- tralized mode of international television coproduction that involved in- stantaneous performance, translation, switching, and transmission. *Our World*'s status as a "live" broadcast was somewhat ironic, however, since it required two years of international collaboration, preproduction planning, and technical preparations.

One of the most important structures established in this early satellite broadcast is an imaginary construct or Western fantasy I will call "global presence." As Jeffrey Sconce explains, the concept of electronic presence dates back at least to the nineteenth century and has been variously de- scribed over the years as " 'simultaneity,' 'instantaneity', 'immediacy,' 'now- ness,' 'present-ness,' 'intimacy,' 'the time of the now.' " As Sconce suggests, "this animating, at times occult, sense of 'liveness' is clearly an impor- tant component in understanding electronic media's technological, tex- tual, and critical histories."[4] In this chapter I develop the term *global pres- ence* to historicize the meanings of "liveness" or "presence" in the context of satellite and television convergence in the 1960s. During this time the meanings of "liveness" and "presence" were indistinguishable from West- ern discourses of modernization, which classified societies as traditional or modern, called for urbanization and literacy in the developing world, and envisioned mass media as agents of social control and economic lib- eralization.[5] Emanating from Western nation-states, the satellite spectacu- lars were imagined as the cutting edge of the modern, the most current or present form of cultural expression. The end point of modernization, then, was constructed as the capacity to be technologically and culturally integrated within a new system of global satellite exchange. Developing nations could only claim themselves as "modern" if they were in range of American, Western European, or Japanese satellite television signals, earth stations, or networks.

Setting the Global Stage

Telstar and *Early Bird* linked the United States and Europe, and *Syncom 2* connected the United States and Japan, but as industry executives foresaw the "swelling global audience" of "space-age TV," they sought to develop programming that was more fully global in reach.[6] Given their technical successes in the early 1960s, broadcasters were ready to take on a bigger challenge. As ABC's James C. Hagerty predicted, live satellite transmission from abroad would be limited almost entirely to "great human events — a coronation, a summit meeting, a sports event." He continued: "As for entertainment, the consensus is that after the novelty of Bob Hope live from London's Palladium wears off, such shows will be no factor. Furthermore, the time zone differences eliminate mass audiences most of the time."[7] Broadcasters schemed to develop cultural events appropriate for live international satellite transmission after a series of experiments via *Echo, Telstar, Syncom*, and *Early Bird* that took place from 1960 to 1965.

Our World was conceived in 1965 by a handful of producers from the BBC's TV Features and Science Departments. Aubrey Singer, the BBC producer of the transatlantic satellite relay *The Town Meeting of the World* (1965), spearheaded what was initially called the "Round the World" project, gaining the support of the European Broadcasting Union and traveling to different countries to assess the availability of technical facilities and broadcasters' interest. The U.S. commercial networks shied away from the project and left it in the hands of the National Educational Television (NET) network (which became PBS in 1967). In September 1966 representatives from eighteen nations met in Geneva to discuss the program's development.[8] At this meeting participants agreed that the program would have no political content, that no item would be included without full knowledge of all participants, and that the entire program would be live.[9] This meant that *Our World* would differ from earlier satellite relays, which tended to focus on the activities of political and corporate officials and feature post-

cardlike vistas of timeless historical monuments in the United States and

Western Europe.

Our World emerged amid important international discussions about the regulation of satellite communication. By 1967 a live international television program that not only linked the East and West but also North and South was both feasible and desirable, since many of the participants were also UN members who encouraged uses of space technologies that would "benefit all of mankind." In 1963 the United Nations General Assembly unanimously adopted the first of several outer space treaties, which provided that "outer space and celestial bodies are free for exploration and use by all states in conformity with international law and are not subject to national appropriation."[10] To encourage further international cooperation in this area, UNESCO convened a special meeting of experts from around the world in December 1965. Scholars, engineers, political officials, and broadcasters were asked to advise on a long-term program "to promote the use of space communication as a medium for the free flow of information, the spread of education and wider international cultural exchanges."[11]

While the producers of *Our World* did not participate directly in the meeting, the discussions shed light on the various ways world leaders imagined life in the age of the satellite. Taking satellite access almost for granted, Western leaders were primarily concerned with shifts in lifestyle. Stanford Professor Wilbur Schramm predicted that "the pace of living in the satellite age may require man to learn how to get along with less sleep, or at least to organize his working and sleeping hours so that they coincide better with time schedules in other parts of the world that most concern him."[12] The English broadcaster Lord Francis Williams suggested that with satellites "the opportunity . . . will exist for ordinary men and women to participate directly as observers in every event of public importance in the world as it actually takes place and with the same immediacy as if they were physically present."[13] And Arthur C. Clarke described satellites as the "nodal points" in the "nervous system of mankind," predicting an age in which they would "enable the consciousness of our grandchildren to flicker like lightning back and forth across the face of this planet. They will be able to go anywhere and meet anyone, at any time, without stirring from their homes . . . all the museums."[14] Each of these comments conjures up a world with the Western individual smack at its center, keeping track of "areas that most concern him," observing firsthand "events of public importance in the world," or having the capacity to "go anywhere . . . without stirring from . . . home." Such comments reinforced a fantasy of global presence in which the world is figured as a realm of access and familiarity.

Leaders from the Soviet Union who attended the meeting had a different perspective. University of Moscow Professor N. I. Tchistiakov emphasized the "equal right of participation of all parts of the world and all countries . . . to balance the powerful flow of broadcasting and information from developed countries by an equal flow from developing countries."[15] Leaders from third world nations such as Pakistan, Nigeria, and India insisted on subsidized access to satellites for underdeveloped nations and proposed that satellites be used in education initiatives throughout the developing world. Nigeria's I. O. A. Lasode proposed that a ground station be built in Nigeria so the African continent could become part of the global satellite system.[16] The Pakistani engineer M. M. Khatib proposed that engineers and scientists from Asia, Africa, and Latin America should be included in the stages of satellite experimentation, trial, and observation so they could acquire technical knowledge and a sense of belonging to the world's satellite development group. This, he believed, would make the global satellite system more genuinely a "world community project."[17]

In 1967, the same year *Our World* was produced, UN members signed the Outer Space Treaty, which provided for free use of outer space in accordance with international law, prohibited national appropriation of outer space, and made states the sole responsible entities for observing and enforcing its provisions.[18] Even before this treaty was signed, however, the United States and the Soviet Union had been appropriating outer space for national security, deploying top-secret satellite espionage systems into orbit. In addition, the United States had been actively working to commercialize the global satellite system since the early 1960s, when it had formed two public corporations: COMSAT (the Communications Satellite Corporation) in 1962 and INTELSAT (the International Telecommunications Satellite Consortium) in 1964. Although both were public corporations, mandated to operate in the public interest (COMSAT) and to promote "world peace and understanding" (INTELSAT), they clearly were designed to benefit the U.S. economy first and foremost.[19]

Gestures toward international cooperation in the development, regulation, and use of satellites were often influenced by cold war politics. On
June 21, 1967, four days before *Our World* was scheduled for relay, the
Soviet Union announced its withdrawal from the broadcast based on its
belief that the United States, England, and West Germany were supporting Israeli aggression in the Middle East, compromising the program's
humanitarian aim.[20] Following the Soviets' lead, the other Eastern bloc participants—Poland, Hungary, East Germany, and Czechoslovakia—with-
drew. Producers quickly added Denmark to *Our World*'s roster and ended
up with fourteen rather than eighteen contributing countries and beamed
the show's signal to viewers in twenty-four rather than thirty nations. The
communist bloc's withdrawal from *Our World* demonstrated the use of liveness for an overt political purpose—that is, to call attention to what the
Soviets perceived as inappropriate Western intervention in the Six-Day War
in the Middle East. As one Soviet leader declared, "The radio and television
organizations of USA, England and the Federal Republic of Germany . . .
are engaged in a slanderous campaign against the Arab countries and the
peaceful policy of . . . socialist states."[21] The conspicuous absence of the
communist nations on the day of the broadcast (especially since for months
promoters had highlighted their participation) complicated the "globalist"
claims of the show. But despite the last-minute cancellation of the USSR and
its allies, *Our World* aired as scheduled.

Since many of *Our World*'s organizers either emerged from or supported
the BBC tradition of public service broadcasting and were aware of UN discussions about the use of satellites in the interests of all humankind, they
agreed that the program should have a humanitarian theme. At a meeting in January 1967 representatives of the participating nations agreed to
develop the show's theme around the "population explosion" because it
was deemed "equally valid and important to people all over the world."[22]
During the 1960s, population growth was declared a global crisis by Western sociologists, economists, biologists, anthropologists, and geographers.
Books such as *The Population Explosion* (1962), *The Population Dilemma*
(1963, 1969), *The Silent Explosion* (1965), and *The Population Bomb* (1968),
to name a few, likened population growth in the developing world to a ticking time bomb that threatened to wreak havoc worldwide.[23] As Paul Ehrlich
explained in his widely read *The Population Bomb*, "each year food produc-

tion in undeveloped countries falls a bit further behind the burgeoning population growth."[24] If population control measures were not instituted immediately, he argued, people in developing countries faced mass starvation.[25] The book's cover depicted a bomb with a short fuse above a panicky catch line: "While you are reading these words four people have died from starvation. Most of them children."

Our World's producers perceived the live satellite broadcast as a unique way to publicize and visualize what they believed was an urgent global crisis. Though they chose population control as a way to build bridges, the theme divided the world once again. Whereas the Soviet withdrawal highlighted political tensions between the communist East and the democratic West, the population explosion reinforced divisions between the industrialized North and the underdeveloped South. As Ehrlich insisted, the world's countries "can be divided rather neatly into two groups: those with rapid growth rates and those with relatively slow growth rates."[26] And as the *Our World* script bluntly put it, the "growth rate is not equal all over the world; because in a sense our world is two worlds. If you are in reach of this programme, you almost certainly belong to the industrialised world." If developing countries did not even have the science and technology for birth control or efficient forms of agriculture, so the logic went, how could they ever participate in such a live satellite television event?

Since *Our World*'s remote cameras did not venture into third world countries, the population problem was visualized as a series of statistics, graphics, and prerecorded images of "hungry people." Its population control theme said as much about the Western imaginary as it did about third world living conditions. While producers may have had good intentions, the absence of both the Eastern bloc and developing countries within the show revealed its self-promotion as a "globe-encircling-now" to be somewhat of a farce. The "global" scope of *Our World* was particularly problematic given that Nigeria, Pakistan, and India had expressed a desire to participate in such "world community projects" during the UNESCO meeting of 1965.[27]

Popular intellectuals such as Arthur C. Clarke and Marshall McLuhan insisted satellites would flatten social hierarchies and unite people across the planet in a "room-sized world" or a "global village." But such metaphors concealed the ways in which live satellite broadcasts were being used to reassert Western hegemony during a period of spatial flux—that is, during a period of decolonization, outer space exploration, and cold war geopolitics. McLuhan and Fiore wrote in 1967, "Ours is a brand new world of allatoncenness. 'Time' has ceased, 'space' has vanished. We now live in a global village . . . a simultaneous happening."²⁸ Rather than reiterate Western liberal ideals of world unity, however, I use the term *global presence* to challenge and destabilize the global village metaphor by exposing neocolonial strategies at work in the "liveness" of early satellite spectaculars such as *Our World*.

Satellite Televisuality

Our World combined the conventions of the film newsreel, the travelogue, and the television variety show. It was divided into five major segments that emphasized problems faced by, and excellence within, the global community.²⁹ In the opening sequence the words *Our World* appear in the frame, and the show's title is announced and shown in several different languages. The BBC host enters a set furnished in minimalist, otherworldly space decor, and as satellite images of a cloud-covered Earth are projected on a seventy-meter-wide screen behind him, he proclaims, "Twenty thousand miles up in the sky, satellites are beaming these pictures into millions upon millions of homes, and viewers in 24 countries all around the world are at this moment watching them together."³⁰

Viewers then witness four "live" child births from maternity wards in Sapporo, Japan; Mexico City, Mexico; Edmonton, Canada; and Arhus, Denmark—representing both hopes for an egalitarian future and Western anxieties of an impending "population explosion." Remarkably, producers planned months in advance to represent Mundo, the Mexican baby, as premature, describing him as "fresh, red and still unseparated from the umbilical." And the baby from Edmonton was a Cree Indian. Since two of the four babies were of color, their births were used to dramatize the population problem. As the babies are "delivered" live via satellite into the studio, the narrator explains, "We can represent the crowded, expanding world by charts and maps and symbols, but none of us can ever see it, at least not as a whole, not as one great family circle as we are at this moment." In short, the world's "great family circle" included only nations that could uplink

Our World featured babies from different countries born live on global satellite television. Courtesy of the European Broadcasting Union.

and downlink with the industrial West. Still, the babies are offered as an expression of universal diversity and oneness, framed in a mosaic, as the narrator proclaims, "[Four] babies. Only [four] out of some eighteen hundred born in the few minutes since this programme began. [Four] whose lives are likely to be worlds apart: born at the moment in history when it is first possible to see round the planet in a moment of time."[31]

The program's narrative was structured around the lives of these four babies literally born into a world of live satellite television. The announcer asks, "What sort of world have they come into? What are people up to around the globe on this June evening in the late 1960s?"[32] These questions motivate the transition to "This Moment's World," a segment designed to immerse viewers in a "panoramic look at people and their activities at a precise moment in various parts of the globe."[33] It begins with a wide shot of Earth—a "full disk" view impossible before the installation of *ATS-1*. Emphasizing the unique vantage point of the remote sensing satellite, the announcer explains, "This is our world as no one *on* the world can see it. Somewhere on this indistinct circle, over three billion people are working, playing or sleeping—or watching this picture."[34] Viewers then move from

the orbital perspective of the satellite to take a whirlwind tour around the globe, stopping at a traffic jam in Paris, a steel mill in Linz, a weather station on Mt. Fuji. Although it violated their apolitical intentions, producers added a last-minute visit to Glassboro, New Jersey, where President Johnson and Premier Kosygin were meeting to discuss world peace. Cameras revealed a horde of international television crews and crowds of protestors outside the meeting holding large signs reminding the leaders, "Peace Depends on You Two."[35] This political pit stop reminds us that live satellite television not only captures but also defines events of global significance by virtue of where its remote cameras land.

"The Hungry World" segment exploits the population explosion theme, opening with a prerecorded photomontage of masses of "hungry people" gazing into the camera as the announcer assures viewers, "We are doing something to help them. Around our world scientists are searching urgently for new means of feeding the ever-growing numbers of mouths." Rather than directly addressing the problem of world hunger (for instance, by critiquing the unequal distribution of resources), the segment focuses on "scientific and technological advances" designed to intensify food production. Viewers witness Ron Caldwell's "convoy of high tech [farm] machinery" in Wisconsin, a hyperproductive shrimp farm in Japan, and an algae lab in Canada. Ironically, the segment spotlights efforts to get species like shrimp and algae to overproduce unnaturally so that the proliferating human population can sustain itself.

"The Crowded World" segment reinforces the theme of overpopulation by proclaiming, "The sheer crowding together of people is an even greater threat to the quality of our children's lives. . . . If we go on growing at the same rate . . . the human race has only 450 years left . . . before extinction by proliferation. Our cameras could not reach the hungry world, but the crowded world is all around them." Viewers see helicopter perspectives of New York City skylines and crowds of Muslim worshippers and shoppers in Tunis. The segment also features housing plans designed to alleviate the pressures of overpopulation, such as Montreal's Habitat (a prefab apartment complex with the "most modern rooms in the world") and Scotland's Cumbernauld, a "visionary [suburban] town" where, the announcer explains, "you are free of the monster"—a reference to the "slums" of Glasgow only fourteen miles away, one of the unspoken side effects of industrialization.

The next segment, "Aspiration to Excellence," celebrates the physical and artistic talents of the industrialized world. This athletic segment sets out to reveal that "even in those parts of the world where men have conquered the basic problems of food and housing, striving does not stop. They are always trying to do something better and better."[36] Viewers watch fifteen-year-old Canadian swimmer Elaine Tanner try to beat her own one-hundred-meter butterfly world record, Swedes plunge through treacherous white water in tipping canoes, and the Italian D'Inzeo brothers jump over intimidating fences on champion horses. Another segment, "Artistic Excellence," profiles artists "whose work," we are told, "is our own pleasure," including Franco Zeffirelli directing a scene from *Romeo and Juliet* in Italy, Miro and Calder at work in the South of France, the opera *Lohengrin* in rehearsal at Bayreuth, pianists Van Cliburn and Bernstein rehearsing at Lincoln Center, Mexico City folk dancers, and the Beatles recording "All You Need Is Love" in a London studio. While the program positions people of the developing world as struggling for basic biological sustenance, Westerners are portrayed with the leisure to engage in abstract intellectual and aesthetic pursuits and to enjoy the body as an expression of individual excellence. In this way the program masks class differences within industrialized nations by contrasting a culturally and economically *pro*ductive West with an essentially *re*productive third world.

The program closes with "The World Beyond," pushing the narrative of Western scientific and technological progress into outer space, where, viewers are told, "man pushes forward the outer limits of his knowledge of our world and those beyond."[37] For the first time, cameras brought viewers to Cape Kennedy's Moonport and to an Australian astronomical observatory. Through a radio telescope in Parkes, Australia, viewers "took a voyage to the limits of the universe—a trip to the edge of time." As the camera panned across their faces, a group of scientists huddled around their instruments and interpreted for the world their data about the earth's interstellar origins. This final segment of *Our World* brings its claims to global presence full circle, for the power to see and experience the earth as a unified totality brings with it the power to know and contextualize the relations of those dwelling on it.

Producers promoted *Our World* as a unique televisual experience, as "the

first time in history that man [could] . . . see his planet as a single place
in both time and space."[38] Not only did satellite vantage points and carto-
graphic perspectives represent the "whole earth" as a unified object via the
Earth shot, but a range of stylistic strategies were deployed and combined
to construct the show's discourse of global presence. Indeed, the show initiates practices that have become the hallmarks of live global coverage. If,
as Caldwell suggests, televisuality refers to the stylistic excesses and self-
exhibitionism of the medium, then *Our World* marks a key moment in this
history. For it reveals television pushing its own limits—extending itself
technologically, ideologically, culturally, and economically as a global system of seeing and knowing. To reinforce this point, I will describe four
practices that emerged in *Our World* and can be understood as part of an
ongoing mode of live satellite television production: spotlighting the apparatus, spatial relations of global presence, scheduled or canned liveness,
and time zoning.

Spotlighting the Apparatus

Our World constructs its global presence by constantly calling attention
to the mode of its production as a spectacle. This mode of live international television production is made possible by a technical infrastructure
that includes satellites, ground stations, signal converters, control rooms,
studios, remote cameras, microwave links, cables, phone lines, and receivers. To prepare broadcasters for the program, producers distributed
flow charts that diagrammed the technical infrastructure supporting the
show, and stylized versions of this preproduction document made their
way to the screen. *Our World*'s liveness is organized in part as the visualization of the signal's real-time generation and movement through this infrastructure. The show, in other words, displays and maps the trajectory of the
signal as it takes shape and moves from place to place. During the show
cameras and control rooms are featured as part of the mise-en-scène, and
narration is used to celebrate the wondrous mode of live global television
production. In the show's first few minutes the narrator invites viewers
on a "journey round the globe through a network of landlines and microwave links and ground stations and satellites." He continues, "in fifty-three

control rooms all round the world, production teams are monitoring and selecting the hundreds of pictures and sounds from five continents which will combine to make this historic program." The experimental nature of the broadcast was exploited to create suspense, and during the transition from Tokyo to Melbourne the narrator highlights the extensive yet delicate infrastructure, warning viewers, "This is the most difficult technical switch!" Here liveness is constructed through the spectacle of the working technical apparatus (which is always haunted by the equally exciting possibility of technical failure).

Our World would have been impossible without satellite technology, yet despite the show's continual celebration of other parts of the technical apparatus, the satellite remained invisible. Satellite technology functions as a structuring absence. Since the orbiting satellite itself could not be represented "live" (at least not in 1967), it is recoded in the broadcast as other space-age stuff. It emerges, for instance, in the otherworldliness of the set design and in the unmotivated electronic noise emanating from "out there."

The spotlighting of the apparatus is connected to what Thomas Elsaesser identifies as television's "standby mode," which involves "a self-staging of television technology and power, as the whole hardware infrastructure of satellite hookups, equipment-laden camera crews, frontline reporters in Land Rovers, and telephone links via laptops becomes visible and audible, making time palpable and distance opaque."[39] Elsaesser links this mode to global television networks such as CNN, but these practices emerge even earlier in the 1960s satellite spectaculars like *Our World*, which gave expression to and dramatized television's standby mode. The phrase "spotlighting the apparatus" refers to the stylistic devices by which television foregrounds the means of its global production, showcasing the dispersed machinery that makes live international transmission and reception possible. Such stylistic flourishes are especially pronounced when technologies converge as they are often imagined as expanding, extending, or overtaking the capacities of another.

Spatial Relations of Global Presence

Our World's fantasy of global presence is also encoded through a series of spatial relations between studio (or technological), global (or geographic),

The *Our World* set was
designed as an otherworldly
space both connected to
and detached from the earth.
Courtesy of the European
Broadcasting Union.

and remote (or performative) sites. In this mode of satellite television the studio plays a central function as the repository for and regulator of remote feeds coming in via satellite and landlines from around the world. Rather than being a static space of talking-head narration, *Our World*'s studio is constructed as a spectacular space of multiple screens, dynamic lighting, and high-performance technical activity. Space-age sounds, spotted patterns of floor light, and minimalist design portray the studio as an unearthly or otherworldly space, and the show itself is presented as a deterritorialized set of fast-moving signals zipping across the planet via satellite.

In the studio the global is represented as a series of simulations: as a fourteen-foot 3-D model of Earth, as screen projections arranged to convey the planet's "infinite variety," and as television signals being instantly generated and transmitted. One sequence dramatizes this last form of simulation explicitly when a cameraman emerges from stage right, dollies across the set, and points his camera at the enormous model of Earth dangling from the ceiling. The cameraman's Earth shot is instantaneously projected on a wide screen behind him and quickly transforms into several smaller frames that feature the "achievements of man" arriving to the studio from remote locations. This sequence spotlights the technical apparatus while encouraging the viewer to see and understand the correspondence between the camera (boldly marked with the show's logo) and the "live images" it delivers to the screen. In this sense the television studio is constructed as a portal to the world "out there," filtering feeds that pour in and packaging them for the screen. The studio works to harness or interiorize the global (and the orbital) within the rubric of the televisual. The fantasy of global presence is predicated on an imagining of the TV studio as simultaneously connected to and detached from the world: It assumes an orbital position distant enough to visualize and construct the world as a "whole sphere" while remaining instantly within reach of its most remote parts.

Geographic maps are used to transition between studio and remote views, orienting the viewer spatially and symbolically "grounding" the ethereal (or, more appropriately, the orbital) signal within an earthly field of representation.[40] During switches between distant places (say, Melbourne and Paris) a world map appears, and the signal's trajectory is plotted from one point to another. "This Moment's World," for instance, intercut feeds of a Tunis camel driver listening to *Our World* on his transistor radio; aerial images of boats at Huelva; a roundup in Ghost Lake, Alberta; bikini-clad girls on the beach in Santa Monica; and road construction in Tokyo. The segment could only portray the show's "full circle round the world" using maps to signify movements from one part of the planet to another. This visual strategy also aimed to dramatize the concept of the "global village." As producers stated in a press release, "For two hours this afternoon the world will take a step closer to Marshall McLuhan's concept of a 'global village' as an estimated 500 million persons in 30 nations witness the first globe-girdling telecast in history."[41]

As geographic maps and plotted trajectories usher in this "globe-girdling" form of television, "live" remote feeds fill the frame, and local narrators introduce activities or performances in their native tongue. These remote views are not static panoramas or tableaux; the cameras are extremely mobile, showing multiple perspectives of the same event and reinforcing the show's claim to global presence. Producers encouraged moving rather than static perspectives, insisting that the show be concerned with people "for humanity is of the moment, whilst buildings and natural scenery are relatively timeless, static, and tend to make uninteresting television."[42] On the set of Zeffirelli's film *Romeo and Juliet* in Tuscany,

Italy, the camera lurks through the rehearsal as if a spy, following the cast and crew as they move through blocking points in an Umbrian church and zooming in to Zeffirelli's face as he passionately directs a scene. To feature a traffic jam in Paris, producers intercut mobile helicopter perspectives with those of several cameras stationed on intersecting freeways. Thus the spatial relations of global presence not only were organized around the alternation of studio, geographic, and remote views but also were constructed through the camera's excessive mobility. Combined, these stylistic strategies worked to naturalize television's presence as part of earthly space itself.

Just as a hierarchy of discourse exists in television news, a hierarchy of liveness exists in the spatial relations of global presence.[43] The live remote feed is the most volatile view, the most unpredictable site of the televisual. As a result, it is never able to emerge raw, untended, unanchored. Put another way, the live remote signal is never seen; it is always screened, emerging through the map or the studio or being "anchored" by narration. The potential for the live-via-satellite view to reveal something unplanned or unstaged has historically been carefully regulated and contained. In the remote live-via-satellite view, then, we do not see the world simply unfolding so much as we see the process by which everyday matter and motion have been packaged to support the Western fantasy of global presence.

Scheduled or Canned Liveness

Although *Our World* constantly called attention to itself as "live," specifically as the "miracle of a globe-encircling-now," like other forms of television it was carefully scripted, meticulously planned, and ardently rehearsed. What distinguishes it from other television formats, which are often characterized by their repetitious structure, is its status as a unique occasion—a "spectacular." As such it anticipates the age of the global media event—an age in which the world's "liveness" is not only packaged but *scheduled* for maximum visibility.[44] While there were many live shows "scheduled" during the 1950s, *Our World* was one of the first to expose the scrupulous *scheduling of liveness.* In this case the scheduling of liveness involved buying or negotiating for time on several satellites, preempting national and local television schedules, arranging and timing each of the contributed sequences, programming the broadcast at a time when it could be seen simultaneously by the maximum number of viewers in different parts of the world, synchronizing the image with multilingual sound tracks, and ensuring that the program aired on time. In other words, *Our World*'s liveness was more a negotiation of preexisting temporal structures than a magically unfolding actuality, a "globe-encircling-now."

Another way to think about this is to imagine *Our World*'s liveness as "canned," much like the laughter on a sitcom sound track. Just as the guttural dimension of the laugh has been harnessed by the televisual apparatus and transformed into the generic signature of the sitcom, the unpredictable volatility of remote occurrences becomes part of television's scheduled annihilation of time and space. While the sitcom promotes itself with a laughter that comes from nowhere, *Our World* celebrates itself as offering a liveness that comes from elsewhere. Still, a lurking tension related to the fantasy of authentic liveness remains in the live satellite broadcast, which is ultimately related to the possibility of technical failure. The more that *Our World* exposed its infrastructure, the more pronounced became the question of whether or not it would work as scheduled. Indeed, the moment of transmission (especially of elaborate experimental broadcasts) is always underpinned by this question. But in *Our World* even the breakdowns were planned: The script included detailed instructions about what to do in the case of technical failures. It is the moment of technical breakdown that in fact takes us closest to gratifying the fantasy of liveness—a fantasy affirmed when we see the apparatus itself resisting its own timing, failing to conform to its own schedule. This is when the medium's presence is made most manifest. In *Our World* there were a few noticeable delays in the transmission when translation or local narration did not start on cue or when the image went black or fuzzy. Such gaps most indulge the fantasy of television's immediacy, for in these moments we become most aware of how television's timing organizes the world.

Our World's most pronounced instance of "canned liveness" occurs in the Mexican cultural segment. Producers' concerns about Mexico's ability to deliver its remote feed meant the segment had to be prerecorded. It was presented as "live" but was actually the only prerecorded remote feed in the two-hour broadcast. The sequence featured Mexican folk dancers and singers performing in the city and countryside. The performances, which included flashy skirt-twirlers, flying doves, cavorting cowboys, and

singing senoritas, were repeatedly intercut with an image of two Mexican technicians watching what was presumably the videotaped version of
the segment as it moved through a reel-to-reel player and out into the
world. The performances were also juxtaposed with an image of several
female performers huddled around a television monitor on a city park lawn
watching themselves as part of *Our World*. The repeated representation of
the scenes of transmission and reception are important because they construct Mexico, one of only two developing nation participants, as part of
the "global present"—even though its contribution had been recorded and
edited days in advance so it could be transmitted on cue.

In the Mexican segment liveness is articulated with Western discourses
of modernization and development. Not only does the sequence spotlight
the apparatus, showing Mexican technicians surrounded by equipment
in a control room, but it quite literally generates an image of the nation
"developing" as a set of signals. Producers sent instructions to local directors encouraging them to exploit the "developing element" of events
whenever possible.[45] Since Mexico's performances had been prerecorded,
the so-called developing element was expressed as Mexico's very capacity
to transmit and receive a satellite television signal, which became a symptom of its modernity. Indeed, in this era of satellite spectaculars modern nationhood increasingly hinged on the capacity to achieve global presence—that is, to have the technical facilities and knowledge to uplink and
downlink with the flows of a global media economy. Mexico's inclusion in
Our World was particularly significant since it would host and relay live television coverage of the Olympic Games to the world via satellite a year later,
in 1968.

Just as development discourses were applied to signals emanating from
non-Western nations, canned liveness worked to mitigate the potential for
unscripted action or technical breakdowns to come from developing nations during live satellite transmission. It differentiated a core of established European broadcasters from more peripheral and tenuous contribu-
tors, especially Mexico, which was subtly displaced from the "global now"
by virtue of its prerecorded segment. *Our World* produced a global mapping
of technological development, dividing the world into zones of technological progress and illiteracy, and used the liveness of the satellite-relayed

signal to dramatize and reinforce them. Those in reach of the satellite spectacular were, by this logic, already developed or developing, and those beyond it remained in a hinterland of underdevelopment.

Time Zoning

As implied by the notion of "scheduled liveness," global presence is also
articulated through temporal relations. I invoke the phrase "time zoning"
here not only to refer to the boundaries of Greenwich time but also as a
metaphor for the ways that multiple temporalities (or time-based imaginaries) intersect in live satellite broadcasts. One of the most frequent and direct
ways the show constructs its global presence is by referring to itself as a
"globe-encircling-now." Arjun Appadurai critiques the notion of a "global
now" as reducing scattered and diverse lived world experiences into a Western expression of global modernity. As he suggests, Western intellectuals have "steadily reinforced the sense of some single moment—call it the
modern moment—that by its appearance creates a dramatic and unprecedented break between past and present."[46] But as much as *Our World* tries
to establish a singular simultaneity or a "global now," it can do so only by
interweaving various time-based imaginaries, which I refer to as a practice
of *time zoning*.

Our World's narrative conflates such trajectories as the life spans of newborn babies, the modernization of the developing world, the history of
Western civilization, the evolution of Earth as an astronomical body, and
the rate of population growth. The "global now" is articulated not so much
as a definitive present but rather as a zone of multiple temporalities—a
zone in which various time-based imaginaries are assimilated, combined,
layered, reordered, and rearticulated, but a zone that the West struggles to
control as *one* "time of the now." As Appadurai notes, "We cannot simplify
matters by imagining that the global is to space what modern is to time. For
many societies, modernity is an elsewhere, just as the global is a temporal
wave that must be encountered in *their* present."[47]

At the most literal level *Our World* constructs itself as a "globe-
encircling-now" by virtue of its crossing of the boundaries of Greenwich
time. In the same way that viewers are oriented in global space with cartography, they are oriented in global time with the Greenwich clock, which

was keyed over maps to indicate shifts in time with changes in place. The show emphasizes shifts in local time zones only to highlight how smoothly they can be traversed by satellite television. One promotion pitched the show as an experience in televisual time travel, boasting, "Time differences around the world, and the ability to switch rapidly through the magic of space-age electronics, will allow cameras to take viewers from 'now,' back to 'yesterday,' and ahead to 'tomorrow.'"[48]

To celebrate the speed of satellite transmission, one animated sequence stages the movement of dawn around the earth with special lighting effects and compares it to television signals beamed from one continent to another via satellite. The announcer explains: "The sun lights up only half the globe. But television can beat the sun. Our cameras can be where it is noon and midnight, dawn and sunset, summer and winter, today and tomorrow, and all at the press of a button. The dawn creeps around the equator at a mere thousand miles an hour, but our pictures flash around at a hundred and eighty six thousand miles a second."[49] Satellite television is imagined as passing through the world's time zones and seasons, as having the capacity to outrun the earth's orbit around the sun. Symbolically, the sequence implies that the satellite has augured the dawn of a new generation of television technology that is tantamount to solar power itself.

Our World's quixotic celebration of a new generation of satellite television is coupled with a warning about new generations of people. The show employs a series of countdown motifs to dramatize the escalating problem of population growth. In one sequence a montage of extreme close-ups of babies' eyes appears on a large monitor in the studio. The last set of eyes dissolves to a shot of a huge ticking metronome as the announcer proclaims, "They are coming into our world at the rate of three every second. Every click of the Metronome is a new baby . . . 90 a minute, 84,000 a day. The population is increasing at the rate of over half a million every week."[50] The metronomes multiply on the screen and are replaced first by population growth statistics, then by images of "hungry people" gazing directly into the camera. As the segment forecasts a doomful future of overpopulation, it infantilizes third world peoples as objects of paternal pity. But the innocent baby, the program asserts, is also a ticking time bomb that threatens to wreak world havoc.

This segment forcefully brings together the show's most troubling discourses on the "population explosion," echoing the racist and eugenicist undertones of the population control movements of the 1960s and 1970s.[51] Like the Rockefeller Foundation and the Nixon administration's Commission on Population Control and the American Future, *Our World* imagines human reproduction in developing countries as a threat to the economic and cultural supremacy of the industrial West.[52] The show insists that only live satellite television, with its capacity to straddle time zones, link remote places, and package faraway events, is adequately equipped to represent such a global crisis.

In this case the practice of time zoning—that is, the layering of such temporal structures as population growth rates, national development, metronome countdowns, Greenwich time, and broadcast duration—works in the interest of establishing new forms of planetary management and control. As its remote gaze lands on the bodies of "hungry" and "overly reproductive" racialized others, *Our World* constructs satellite television not only as a technology of global presence but of global monitoring, measurement, and classification as well. Such views align satellite television technology with scientific and military practices of observation used to master and control from a distance.[53] The fantasy of global presence is expressed not only through the temporal strategies I have described but also as a set of contradictory impulses to unify the world while dividing it, to traverse boundaries while reinstating them, to assist others while condemning them, and to educate and entertain people while observing them. These impulses are laden throughout the *Our World* text, which plays on the utopian dream of a "global village" only to expose the deep rifts in the Western imaginary that create the need for such a dream in the first place.

Conclusion

Broadcasters went to great lengths to ensure that *Our World* would be understood as a "global" event. After the withdrawal of the Eastern bloc, producers rushed memos to participants encouraging them to continue promoting the show as "global."[54] An *Our World* press release emphasized the liberal humanism that motivated the "electronic miracle" of bringing the world together. It declared "[The show] will . . . inspire men everywhere with the realization that time and distance are no longer effective separators of mankind in this twentieth century, that all men are neighbors in

the same world and that mutual understanding and co-operation are in
the best interests of all."[55] While such rhetoric touted the utopian promise of live satellite television, it was also complicit with neoimperialist discourses that aligned the technology with linear models of development and
modernization and that worked to differentiate the West from the rest of
the world.

Although *Our World* was promoted as a "global" program, its narrative and audiovisual discourses divided the world into North/South and
East/West hemispheres, taking care to distinguish the "free" industrialized world from the impoverished and "hungry" developing world and the
excised communist bloc. After the broadcast an astute *Denver Post* critic
spotlighted the show's exclusionary globalism, noting, "There was a huge
dark area stretching all the way from Japan to mid-Europe. As much as
anything, the darkness dramatized the deep rift that divides the world."[56]
This "dark area" contained second and third world nations, many of which
lacked access to television and/or satellite technologies and remained separated from *Our World*'s historic "globe-encircling-now." *Our World*'s partial
global map served as a stark reminder that the meanings of "development"
and "modernity" were increasingly contingent on access to satellite television technologies that enabled nations to assert their global presence by
participating and representing themselves in live international exchanges.
The modern and developed nation was, according to the show's logic, necessarily one that could readily downlink, transmit, and appear in the live
Our World signal.

Our World was not just a passing moment in international broadcast
history, then. As one of the first instances of satellite and television convergence, it served as a kind of bellwether broadcast that foreshadowed a
range of satellite television practices that would take shape in decades to
come.[57] We can notice, for instance, in this historic moment of satellite
television convergence, two decades before the widespread development of
direct satellite broadcasting services, the early imaginings of a satellite footprint. *Our World* is one of the first times a boundary was drawn across the
earth for the purpose of simultaneous distribution of a television signal.
If we imagined *Our World*'s transmission as a footprint, it would include
Western Europe, North America, and East Asia—which is where the first
direct satellite broadcasting services emerged in the 1980s and 1990s.

Our World also reconfigured the content and form of television with its practice of live international coproduction designed to emphasize the world's diversity and serve as a platform for humanitarianism, albeit from a somewhat problematic Western perspective. The show established patterns of televisual coverage—such as spotlighting the apparatus, spatial relations of global presence, scheduled liveness, and time zoning—that later would be institutionalized by transnational television networks such as CNN, BBC World, and Al-Jazeera during the 1980s and 1990s. As John Caldwell observes, "Satellite broadcasts today seldom emphasize either immediacy or liveness."[58] As satellite use grew more common throughout the 1970s and 1980s, the technology was internalized as part of the televisual apparatus, naturalized as an extension of the medium's far reach, and treated as a technology of signal distribution rather than as one of television's content and form.

Perhaps most important, *Our World* functioned as a key moment in the social construction of satellite television technology: It drew on and combined several of the practices delineated in this book, including live international transmission, remote sensing, and astronomical observation. As it presented television signals ricocheting across the earth, it also showed a "full-disk" view of our planet taken by the ATS-1 satellite and invited viewers to take a "trip to the edge of time" over the shoulders of astronomers at the observatory in Parkes, Australia. Such satellite television practices yielded new forms of visual mobility, generating movements not just across but above and beyond the earth. Through such practices *Our World* highlighted the various discourses that gave shape to early satellite television technology including public education, national development, cultural performance, and scientific observation. In this sense the program signaled the need to reconceptualize the televisual as a site of technological convergence that can activate different discursive modalities as opposed to inevitably functioning as either public broadcasting or commercial entertainment. Satellite television texts might combine scientific observation with cultural performance or public education with military intelligence in the way they produce worldviews and convey knowledges.

Finally, as *Our World* helped to initiate television's global turn, it situated the medium as a technology of knowledge as well. The moment was suggestive of the ways that satellite television would increasingly be used not just to view but also to know and make sense of the planet. But as

Our World offered new visual mobilities, it held Western epistemologies

firmly in place, patrolling lines between the democratic West and communist East, the rich North and the poor South. The show, in effect, constructed the world as alternating zones of productivity and inertia, significance and banality, centrality and irrelevance, technological mastery and incompetence.

Our World might be understood, then, as a transitional moment looking toward the current global system, which Saskia Sassen describes as a "new geography of centrality and marginality." This system is "a very strategic and highly structured geography of transactions and networks" that privileges concentrations of labor and capital between global cities such as New York, London, Tokyo, Mexico City, Paris, Bombay, Taipei, and Sydney, among others.[59] *Our World*'s satellite television practices corroborated just such a system. The program combined remote feeds from many (but not all) of these global cities, flexibly positioning them as networked yet distinct. As *Our World* used satellites to extend international television networks across continents (which oceanic cables had also done), it also streamlined and intensified connections between these select global cities. Finally, as *Our World* visualized this expanding international network infrastructure it adumbrated the global information order of the late twentieth century. The global system that Sassen now refers to as a "cross-border geography" was derived not only through the computer industries of the 1990s and 2000s. This system was also generated through earlier global satellite television practices that worked to isolate, interlink, and consolidate dispersed centers of power, productivity, and authority in the world, while demarcating other territories as beyond the pale.

The community of Alice Springs from the perspective
of the Imparja TV transmitter. Photo by the author.

Satellite Footprints

IMPARJA TV AND POSTCOLONIAL FLOWS IN AUSTRALIA

During the 1970s and 1980s several satellite television experiments took place in developing countries, including the SITE project in India, SACI in Brazil, and Project SHARE in Africa. By relaying educational television programs, the Western sponsors of these projects hoped they would accelerate industrialization processes and help so-called primitive societies, as Arthur C. Clarke put it, make a "quantum leap" into the modern age.[1] These projects reinforced Western discourses of modernization embedded in early spectaculars like *Our World*, positioning the satellite as a technology of national development and global integration. As satellite use expanded to different parts of the world, however, those who programmed and watched satellite television adopted tactics for ameliorating some of its imperializing tendencies and effects. Whereas in 1967 satellite television was used to construct non-Western peoples as an overpopulating mass, by the mid-1980s Aboriginal Australians, who had been excluded from *Our World* even though Australia played a key role within it, developed their own satellite television network called Imparja TV.

Imparja, which means "tracks" or "footprints" in the Aboriginal Arrernte language, is a satellite television network owned and operated by a consortium of Aboriginal media groups, communities, and land councils. The network emerged in 1988 after the Central Australian Aboriginal Media Association (CAAMA) banded with Aboriginal land councils to win a license to operate a transponder on the Australian national satellite, *Aussat*. When it began operations, Imparja's signal was broadcast throughout the Northern Territory and Central Australia. In 1999 a digital conversion process expanded its reach to remote communities throughout the Australian continent and Tasmania.[2] The service relays a mixture of global (mostly U.S.

and British), national (Australian), and local (Aboriginal) television programming to Aboriginal and white viewers scattered throughout the Imparja footprint.

Before Imparja's emergence the late media ethnographer Eric Michaels suggested in his 1986 study "The Aboriginal Invention of Television" that Aboriginal uses of video and television differed so greatly from Western uses that they constituted an altogether different technological invention.[3] Adopting Michaels's logic, I suggest that Imparja constitutes a *reinvention* of satellite television as well. For Imparja TV is unique not only in terms of the indigenous content it distributes but also in its use of the satellite as a means of territorial reclamation and cultural survival in postcolonial Australia. As Arlif Dirlik points out, the term *postcolonial* has various meanings; I invoke it here to refer to "a condition in formerly colonial societies," as well as to "the epistemologic and psychic orientations that are products of those conditions."[4] Where development projects used satellites in an effort to "modernize" rural populations in developing countries such as India and Brazil, Imparja TV represents an appropriation of an Australian national satellite to support postcolonial Aboriginal struggles for territory, cultural autonomy, and social equality.

Since Imparja operates with a very small, state-subsidized budget, the bulk of its programming is derived from inexpensive Australian, British, and American satellite television feeds. But despite its heavy use of Western syndicated programs, Imparja is also the only Australian television service that regularly relays native language news, entertainment, and educational programming made by Aboriginal people. While a satellite network such as Rupert Murdoch's Star TV targets a full third of the world's population, Imparja's vast footprint reaches only 3 percent of Australia's population—fewer than two hundred thousand people.[5] Like Star TV, Imparja covers a broad territory, but it does so in order to reach a tiny audience. Imparja TV is significant, however, for it has enabled Aboriginal Australians—a social group that has experienced territorial colonization, cultural invasion, and anthropological scrutiny—to reestablish ties across vast territories and regulate the flow of foreign TV signals coming into Aboriginal lands.[6]

In this chapter I use the case of Imparja TV to rethink the concepts of television flow and the satellite footprint. There has been so much Western scholarship on Aboriginal cultures that this discussion runs the risk of reinforcing a Western fascination with them. In an effort to try and avoid such positioning, I have imagined this chapter as an exercise in reversal to the extent that I hope the analysis of Aboriginal satellite television practices will inform, alter, and reshape the way Western television critics understand satellite footprints and television flows. Raymond Williams first developed the concept of television flow in 1974 to refer to the organizational structure of U.S. commercial television and to highlight the way commercial interruptions often reinforced dominant ideologies embedded within sequences of television.[7] As Mimi White reminds us, television flow emerges as a "traveling theory," resulting from British scholars' "first encounter" with U.S. television.[8] Few, however, have explored the concept in relation to television's globalization, convergence with new technologies, or use in postcolonial societies. While flow is initially described as a structure of U.S. commercial television, it is important to consider how such structures have migrated to and are reconfigured in television in different parts of the world.[9] As White's observation implies, since flow emerges as a cross-cultural discourse, it should be used to make explicit television's relationship to cultural and economic differences within and across national boundaries.

While there has been much discussion of flow, there has been less critical attention directed toward the footprint. *Footprint* is typically understood as a technical term referring to the geographic boundary in which a given satellite's signal falls, but I use the term to describe a cultural territory shaped by the practice of downlinking and uplinking television signals. In other words, satellite footprints are constituted through the distribution of signals, whether raw feeds for television networks or programming packaged for viewers. Multinational media corporations and small television networks alike generate footprints by buying time on communications satellites, selecting and licensing programming from producers or networks, and negotiating with state agencies to establish distribution systems within or across national boundaries. Not just a technical term, *footprinting* is a cultural practice that involves the production of signal territories across sovereign national borders. In the late 1980s and early 1990s the footprinting of Asia and Europe led to problems of "satellite spillovers," which meant a footprint overlapped with the borders of two or more nations, enabling citizens in one country to downlink satellite television signals from another without state authority.[10] The satellite footprint, because it traverses national boundaries, often complicates—and in some cases dislocates—the state's power to regulate television flow. Thus even though the satellite occupies an orbital position, it is deeply embroiled in the reorganization of cultural and political boundaries across the earth's surface. And individuals within satellite footprints are interpellated not only as viewer-consumers who can receive signals with a dish but also as inhabitants of a territory that is simultaneously geophysical and ethereal, bound to Earth and open to what lies beyond it. The footprint, which is constituted by the uplinking and downlinking of signals, is also a zone of linguistic, ethnic, class, or taste-based affinities and differences, bodies and desires, state(s) of jurisdiction and corporate control. The footprints that crisscross the planet are symptomatic of an era of postnational cultural production in which the state's historical role as regulator of television production and distribution is increasingly handled elsewhere. In some instances practices such as censorship, translation, or scheduling, formerly handled by state regulators or public broadcasters, can even be automated and conducted on board the satellite. In short, there is an incredible amount of cultural and political power invested in satellites and the footprints they generate.

Satellite footprints have been mapped on the globe (at least on a wide scale) only during the past two decades, with the emergence of direct satellite broadcasting industries in Europe, Asia, North America, South America, and, most recently, Africa. Many satellite footprints such as those of Star TV, Globovision, and Astra are multinational. But because of Australia's unique geography, Imparja's footprint is subnational. The footprint covers all of the Northern Territory and South Australia, from Bathurst Island north of Darwin, to Port August in South Australia, and in 1999 it was expanded to include "black out" regions, urban areas that had not previously been able to access the signal.

Although Imparja TV controls a satellite footprint, it lacks the financial resources necessary to produce extensive blocks of Aboriginal programming. Imparja has used satellites not only to distribute its signal but also to selectively downlink other networks' flows and reschedule them according to the routines and interests of remote communities. Within these conditions flow is produced through a resourceful practice of filtering and rescheduling material downlinked via satellite. Since much of this material emanates from the United States, England, or urban centers in Australia,

An Imparja TV transmitter.
Photo by the author.

flow is also a site of cultural hybridity that often gives expression to the unequal power relations, historical disjunctures, and economic disparities that shape postcolonial Aboriginal societies. Imparja's footprint and flow are implicated within ongoing Aboriginal struggles for territorial reclamation and cultural survival within Australia and the global economy. As Imparja's Web site proclaims, "Imparja is unique in Australia and the world, as a commercial television enterprise totally owned and controlled by Indigenous peoples. . . . [It] acts as an example to the World's Indigenous peoples of what can be achieved through Aboriginal enterprise."[11]

"*Dallas* in the Outback"

Aboriginal access to satellite and video technologies stemmed from the Australian government's decision to provide Australian Broadcasting Corporation (ABC) programming to viewers in outback territories during the early 1980s. This plan sought to bring white citizens in remote areas into the folds of national culture. In 1982 the Australian government announced plans to license commercial companies to operate four footprints in Australia's remote regions. The Remote Commercial Television Services (RCTS) plan, as it was called, would deliver television programming to viewers in four different footprints. One of the four RCTS footprints—known as the central zone—had a substantial Aboriginal population. The Central Australian Aboriginal Media Association (CAAMA), a group that had already formed an Aboriginal-language radio station called 8KIN, expressed concern about the impact that foreign television might have on the hundreds of Aboriginal communities located throughout the footprint. Many of these communities did not have radio or telephone at the time, and some Aboriginal leaders feared that satellite feeds of Anglo-Australian media would compromise local cultures and languages. Anmanari Nyaningu, a prominent member of the Ernabella community, compared unwanted satellite television signals to land rights violations, stating, "Unimpeded satellite transmission in our communities will be like having hundreds of white-fellas visit without permits everyday."[12] Indeed, the prospect of further white Western invasion—this time in the form of electronic cultural trespassing—motivated Aboriginal communities to wage a battle for control over the central zone's transponder.

By the time the Australian Broadcasting Tribunal invited applications, CAAMA had already organized Aboriginal groups and land councils in an effort to win the central zone license. Aboriginals constituted 40 percent of the total audience within this footprint, and CAAMA insisted that they be directly involved in the development of new satellite television services. In 1985 CAAMA created a commercial company called Imparja Television as the vehicle for license application. Imparja lacked strong financial backing, but it outlined a comprehensive plan to broadcast a wide range of programming that would reflect the diversity of the audience within the central zone RCTS service area. The proposal included Aboriginal language programming produced by and for Aboriginal people.[13] A Darwin-based company called Television Capricornia was the only other applicant. It was well funded and had the support of Darwin television station NTD-8 and the provincial government, but its programming proposals were weak.

Imparja's proposal relied to a certain extent on the aggressive use of a

previously implemented federal program called the Broadcasting in Remote Aboriginal Communities Scheme, or BRACS.[14] In 1988 BRACS provided each of seventy-four remote Aboriginal communities with thirty
thousand dollars worth of media equipment, including a receive-only satellite dish, transmitters for one television and one radio channel, video- and
audiocassette recorders, microphones, cameras, and tripods. The BRACS
program enabled Aboriginals with technical knowledge (which was limited
at the time) not only to selectively access signals off the national satellite
but also to produce and transmit their own programming.[15]

In 1985 the Australian Broadcasting Tribunal held hearings on the proposals in Alice Springs. Aboriginal people from all over the region testified
about the impact that Western programming choices would have on their
cultures and communities. The fact that land councils participated so actively in this battle for the central zone suggests that Aboriginals perceived
their struggle for the footprint as part of their efforts to reclaim control over
their lands. At the time, many Aboriginal communities were involved in
land rights battles resulting from the Aboriginal Land Rights Act of 1976,
which had returned some sacred lands to local control. In the Northern Territory, Aboriginals owned or had claims to almost 50 percent of the land,
and they received modest royalties from mining and tourism. After a long
and heated hearing process, the Australian Broadcasting Tribunal awarded
Imparja a seven-year license that has since been renewed.[16] This license
granted the Aboriginal company rights to deliver radio and television to remote communities in the Northern Territory and South Australia and to
large outback towns such as Alice Springs, Catherine, Tennant Creek, and
Coober Pedy.

The hearings provoked racial tension throughout the region, and many
white residents perceived the matter as a racial public relations campaign
for the federal government. *The Australian Financial Review* reported that
the tribunal's decision "stirred a political and racial hornet's nest in the
north."[17] The Northern Territory's communications minister, Ray Hanra-
han, claimed that the tribunal's decision was "nothing more than a Federal
Government–inspired social experiment in the interest of an activist minority."[18] Many white residents of Alice Springs vowed not to watch Imparja
TV. *The Record* reported that when the Federal Communications Ministry
announced that Imparja was the licensee, it "prompted an uproar by oppo-

sition politicians, who claimed the license was awarded on 'racial lines' and would give a minority group a monopoly over one-third of Australia."[19] Politicians themselves seemed to recognize this battle over the satellite footprint as part of a heated land rights struggle. Despite white protests, Imparja TV, the world's first satellite television channel owned and operated by indigenous people, premiered from Alice Springs in January 1988. The channel reportedly opened with "a sophisticated graphic identifying the station and depicting a boomerang circling the country, accompanied by a futuristic adaptation of Aboriginal music."[20]

Although some Aboriginals resisted the idea of downlinking Western television, financial conditions forced the network to carry inexpensive syndicated programming. Before Imparja won the license, CAAMA director Freda Glynn predicted, "I suppose we will probably have *Dallas* and programs like that during prime time. But in the down time, we will be providing educational and language programs for Aborigines in the area."[21] True to Glynn's prediction, *Dallas* was one of the first series aired on the Imparja channel.[22] Since the network was only partially subsidized by the Aboriginal and Torres Strait Islander Commission (ATSIC) and federal government funds from the Australian Bicentennial Authority, it depended on the advertising revenue generated by Australian and syndicated U.S. and British programs to survive financially. Nonetheless, because Imparja was willing to air shows like *Dallas*, the network became a source for local Aboriginal programming as well.

Only 1 to 2 percent of Imparja's programming is considered "Aboriginal," and most is produced by CAAMA or Imparja. The process of producing and selecting culturally appropriate programming for a population with such diverse linguistic backgrounds has been a great challenge. In the two hundred years since colonization, 90 percent of aboriginal languages had been rendered extinct.[23] Still, during the mid-1980s more than twenty-two Aboriginal languages were still spoken within the central zone footprint.[24] The multilingual *Nganampa Anwernekenehe (Ours)* is one way Imparja has responded to this difficulty. *Nganampa* first aired in 1993 in celebration of the International Year of the World's Indigenous Peoples.[25] Produced by CAAMA, it has been funded by Imparja and ATSIC and operates on a budget of $250,000 per year.[26] *Nganampa* features "dreaming" stories, programs addressing contemporary social issues, and stories on traditional lifestyles gathered from different community-produced tapes.[27] Targeting

Nganampa inscribes
Aboriginal cultural practices
within the televisual.

an Aboriginal audience, the program is produced in languages such as Arrernte, Warlpiri, Pitjantjatjara, and Luritja and is not subtitled.[28]

The dreaming (or "dreamtime") is a Western term that refers to Aboriginal cosmologies.[29] Anthropologist Harvey Arden described the dreaming as "a Genesis-like epoch in which ancestral creator figures—like the Rainbow Snake, the Lightning Brothers, and the Wandjina, or Cloud Beings—traveled on epic adventures across earth's originally featureless topography, giving shape in the course of colossal struggles and battles to each mountain, river, gorge, rock formation, or water hole."[30] Aboriginals characterize the dreaming as "jukurrpa" or "the law," and they retell stories of the dreaming in order to renew and maintain their relationship to the land and their history.[31] The dreaming is not, however, an account of an isolated mythic past. Instead, the dreaming is an ongoing cultural practice by which the land—and Aboriginals' relationship to it—is continually reinvoked in language. Thus, for Aboriginals the landscape is as much a linguistic, cultural phenomenon as it is a physical reality.

As the only regularly scheduled Aboriginal-language television program in Australia, *Nganampa* carries enormous significance for Aboriginals. According to the show's executive producer, "We get a lot of positive feedback about *Nganampa*. Aboriginal people are proud of seeing their culture presented on TV."[32] An on-air promotion for *Nganampa* features a montage of traditional food items, dreamtime paintings, and political demonstrations at which Aboriginals carry signs reading, "Keep Our Law Strong" and "Our Culture Still Lives." Such promotions are designed to inscribe "black pride in what is often a white racist (television) community."[33]

Aboriginal media scholar Marcia Langton claims *Nganampa* is especially significant given that in 1992 "Aboriginal and Islander people were still virtually invisible on the three commercial television networks."[34] Indeed, the ABC and the Australian commercial film industry have a poor record with respect to the representation of Aboriginal people.[35] In 1997 indigenous broadcaster Lola Forrester proclaimed, "I have worked in the media for the past eight years, and can count on my hand the number of times that commercial media has given time, space and vision to positive stories about indigenous Australians. The exception to this practice is in the area of promoting Australia to overseas tourists . . . [but] these images don't portray Aboriginal Australians as part of the modern-day scene; instead they concentrate on how unique the culture is when you compare it to any other culture on the planet."[36] This history of white colonization, social discrimination, and cultural appropriation has led most Aboriginals to contend that it is better to have minimal Aboriginal programming and institutional control over the central zone transponder than to allow another white media company to control the signals beamed down to Aboriginal satellite dishes.

In addition to *Nganampa*, Imparja airs Aboriginal programs such as the BRACS *Show*, which features videos sent to Imparja by various Aboriginal communities; *Corroborree Rock*, an Aboriginal music program that is reportedly "enjoyed by indigenous and non-indigenous alike";[37] and *Yamba*, a children's program about the life of its eponymous red, black, and yellow yerrampe (honey ant), who proudly wears the colors of the Aboriginal flag and speaks *Arrernte*. Imparja TV executives see Yamba as "shar[ing] the richness of his cultural history to his diverse audience."[38] Finally, Imparja produces a nightly news program, *Imparja National News*, that blends global and national feeds with 40 percent to 50 percent of local content.[39] The satellite channel also telecasts an English-language current affairs show called *Urrpeye* (*Messenger*) that explores contemporary Aboriginal issues.[40]

In the 1990s Imparja cancelled *Dallas* and replaced it with U.S. shows such as *Beverly Hills 90210*, *Melrose Place*, *Dawson's Creek*, *The X-Files*, and *Star Trek*. The phrase "*Dallas* in the outback" is significant, however, because it marks the hybrid structures that give shape to Imparja's footprint and flow, and it suggests the need to rethink these terms in more cross-cultural and place-specific ways. While flow may be considered a Western way of programming, the alternation of *Nganampa*, which is fundamen-

Imparja TV production facilities in Alice Springs. Photo by the author.

tally about Aboriginal cultural survival, with national and local commercials and Australian, British, and American feeds and reruns, constructs
it as a site of hybridity. As Lisa Lowe explains, hybridity involves "the formation of cultural objects and practices that are produced by histories of
uneven and unsynthetic power relations. . . . Hybridity . . . does not suggest the assimilation of . . . [minority] practices to dominant forms but instead marks the history of survival within relationships of unequal power
and domination."[41] Imparja's flow is a cultural practice resulting from histories of uneven power relations in postcolonial Australia and the global
media economy. Rather than being assimilated within a Western model
of television organization, Imparja has generated flow that is punctuated
by Aboriginal languages and cultures and selections of foreign content via
satellite. One could argue that flow has always been imagined as a form of
hybridity (Horace Newcomb and Paul Hirsch described it in the early 1980s
as a site of "opposing ideas abutting one another" or "opposing treatments
of the same ideas"),[42] but it has not always been analyzed as such. Pushed
to its extreme, flow analysis demonstrates the pervasiveness of capitalist
ideology within television's structure while revealing the social and economic contradictions that inevitably emerge in a medium that continually
asserts itself as diachronic and globally present. To analyze flow as hybrid

involves exposing the ambivalences and incongruities that emerge in the process of downlinking and arranging Aboriginal, English, American, and Australian shows for broadcast throughout the Imparja footprint.

Satellite Filtering

When Aboriginals first gained access to BRACS equipment, some of them understood the satellite dish as a technology that would allow local leaders to become the gatekeepers of media coming into their communities. A segment in a CAAMA documentary called *Satellite Dreaming* constructs the satellite dish as a shield that would protect Aboriginal communities from further colonization. As adviser Clive Scollay discusses the implementation of BRACS within Aboriginal communities, a graphic sequence dramatizes Aboriginal use of the satellite dish. When a map of Australia appears, three satellites spray the continent with television signals. An Aboriginal man emerges from Central Australia and holds a large shield above his head that enables him to deflect unwanted satellite signals away from the central zone. Indeed, the purpose of BRACS, according to Marcia Langton, was to "allow remote Aboriginal communities to *filter* inappropriate ABC (Australian Broadcast Corporation) programs and insert their own culturally relevant product into the [satellite] service."[43]

Imparja also operates by this logic as it selects Australian programs and syndicated shows from the United States and Great Britain to fill much of its schedule. It is twenty times cheaper for Imparja to purchase and relay blocks of syndicated programming sold by global media conglomerates than it is to produce local, indigenous shows.[44] Still, Imparja's mandate to serve the interests of Aboriginal communities means that its program selections might be determined as much by cultural imperatives as by financial considerations. Imparja employees meet weekly to discuss and evaluate programming decisions and to ensure the network fulfills its mission to "deliver information and communication services to the community, while promoting indigenous culture and values."[45] In 1989, for instance, Imparja's board of directors voted to ban local alcohol advertising to discourage drinking in Aboriginal communities. Since Imparja relied on revenues from national advertisers, however, the network continued to run those commercials but broadcast them in conjunction with a "respon-

sible drinking" public service campaign.[46] As Eric Michaels suggests, "Aboriginal television then is not just the televisual expression of Aboriginal
culture and aspiration. It is also shaped by Aboriginal negotiation of the
existing video and television industry with its mix of local and international
elements, and of the policy (broadcasting and Aboriginal)–government-
funding nexus."[47]

While Imparja provides a first level of filtering by selecting shows that
are thought to be of interest to people in the footprint, this process happens
at a more local level as well.[48] The BRACS plan equipped outstation communities with satellite dishes that enabled them to downlink selectively
from Imparja. Thus, programming may differ within Aboriginal communities, depending on how actively they use their satellite dish and rebroadcast
facilities. While some communities simply rebroadcast the Imparja signal
with little editorial control, using the satellite dish as "a sponge," others
carefully shape which programs will be aired.[49] Yuendumu and Ernabella,
for instance, often preempt Imparja's signal with local dreaming videos,
football matches, or town meetings. But while some Aboriginal communities have been able to produce their own programs, others lacked the technical skill and funding to do so and thus could only receive and retransmit
distant signals of ABC, SBS (Special Broadcasting Service), and Imparja. The
process of filtering remains particularly important within those communities in which local video production is minimal, for it enables Aboriginals
to use the satellite dish itself as a technology of production. Filtering, in
other words, is a low-tech satellite practice of selecting and fashioning a
local stream of television programming from a flood of global and national
signals.

Since the satellite dish is used to regulate which television signals enter
Aboriginal communities, it can be seen here as a technology of national
and/or global rejection as much as one of connection. This is significant,
for it suggests that satellite technology need not be an agent of Western
cultural imperialism but rather can be used to assert local interests and
priorities. Imparja's staff describes its offerings as the result of "cherry-
picking" from national and global satellite feeds.[50] While the metaphor of
cherry picking may resonate with the rhetoric of "choice" used by networks
to promote themselves in an era of multichannel environments, it also
conveys the sense of autonomy derived through Aboriginal use of satel-

Table 1 Sample Imparaja tv schedule (July and August 1999)

Time	Origin	Normal day, August 2, 1999	Naidoc Week, July 11–14, 1999
4:00 p.m.	Imparja	*Yamba*	*Yamba*
4:30 p.m.	Aus	*Pig's Breakfast*	*Pig's Breakfast*
5:00 p.m.	Aus	*Catchphrase*	*Catchphrase*
5:30 p.m.	U.K.	*Neighbours*	*Neighbours*
6:00 p.m.	Imparja	*Imparja National News*	*Naidoc Week Coverage*
6:30 p.m.	U.S.	*A Current Affair*	*A Current Affair*
7:00 p.m.	Aus	*Sale of the Century*	*Sale of the Century*
7:30 p.m.	U.S.	*Veronica's Closet*	*Veronica's Closet*
8:30 p.m.	U.S.	*ER*	*Corroborree Rock / Nganampa* (CAAMA)
9:27 p.m.	Imparja	*Imparja News and Weather Update*	*Imparja News and Weather Update*
9:30 p.m.	U.S.	*Dawson's Creek*	*Dawson's Creek*
10:30 p.m.	Aus	*Good News Week*	*Good News Week*
11:30 p.m.	U.S.	*Melrose Place*	*Melrose Place*
12:30 a.m.	Aus	*Sports Tonight*	*Sports Tonight*
12:58 a.m.	Imparja	*On Track*	*On Track*
1:00 a.m.		Station close	Station close

lite technology. This filtering practice also suggests that the satellite functions as a technology of consumption or even production as much as one of neutral distribution.[51] In other words, Aboriginal "cherry picking" suggests that those who do not have the financial resources to engage in widespread *program* production may devise modes of *flow* production that have more to do with signal selection and scheduling than with the creation of original shows. Thus satellite filtering becomes a practice of producing flow—"cherry picking" programs according to the interests, routines, and rhythms of those within the footprint.

To illustrate this further, I want to analyze a segment of Imparja flow aired during Naidoc Week in July 1999. Naidoc Week is dedicated to the "celebration of Indigenous culture and survival."[52] Imparja adjusted its schedule during Naidoc Week to air more Aboriginal programs and to enable Aboriginal news anchor Catherine Liddle to present a summary of the day's events each night.[53] Imparja and CAAMA both produced Naidoc Week specials, which were rerun later at various times. During Monday night of Naidoc Week, the Aboriginal shows *Corroborree Rock* and *Nganampa* preempted *ER* and were followed by *Dawson's Creek*, a youth-oriented melo-drama set in a white American small town. This sequence is worth de-scribing in some detail because it is suggestive of the hybridities that characterize the network's flow.

The *Corroborree Rock* episode features the January 1999 Survival Day concert with performances by Aboriginal musicians such as the Warumpi Band, Archie Roach and Rubie Hunter, Native Ryme Syndicate, Colored Stone, and Teenage Band Lajamanu.[54] Between performances musicians speak about Aboriginal cultural survival. Addressing Aboriginals through-out Australia, singer Debra Mailman proclaims, "To all the elders of the community across the country, survival day is a day of respecting you and what you've done for our younger generation, for people like me and fight-ing in the war and the struggle you fellas had to give us. Thank you for that. I'll always pay respect to you, our elders." Before his performance Archie Roach explains, "It's only because of them [elders] that I'm here today you know. It's like they were able to . . . survive their time and their struggle so that we can continue. And hopefully there will be a point in time when all their work and all our work will come to fruition." A member of the Ab-original rap band Native Ryme Syndicate says, "The word survivor . . . that's straight out of our black history. We're still here saying it's not history; it's present and we'll still be here tomorrow." The episode is interrupted by ad-vertisements for Telstra's Learn IT program (which shows white children playing the violin), Heinz soup, Austar satellite TV service (an Imparja com-petitor featuring clips of Hollywood movies), the Kmart pharmacy, a local shopping center, Ford compact cars, and Naidoc Week public service an-nouncements. *Corroborree Rock* is followed by a *Nganampa* episode called "This River Still Got Song," which features elders from various communi-ties telling dreaming stories about the Fitzroy River in the Kimberly region, a place where white Australians have been trying to build a dam.[55] Views of the river and storytellers are interrupted by commercials for Ford cars, Head and Shoulders shampoo, Telstra satellite communications, and the Lion's Club Camel Cup Competition. *Nganampa* is followed by *Dawson's Creek*, a youth melodrama set in Capeside, Massachusetts, "a quaint little town that thrives on tourism by summer but becomes all but abandoned during the winter."[56]

Understanding flow as a site of hybridity involves considering how *Cor-*

roborree Rock and *Nganampa* might play in relation to an American series such as *Dawson's Creek* (or *ER* and *Melrose Place* or the UK's *Neighbours* or Australia's *Sale of the Century*, for that matter). The musicians in *Corroborree Rock* transmit their struggle for cultural survival through Imparja's signal, addressing youth and elders throughout the country and the footprint. And Aboriginals living near the Fitzroy River turn their stories into televised narratives that are carried via satellite to viewers throughout the signal territory. Both of the Aboriginal shows are repeatedly interrupted by commercials for satellite television services (Telstra and Austar), compact cars, and Naidoc Week public service announcements, which establish a connection between technologies of mobility and life in Aboriginal communities. This flow segment facilitates Aboriginal claims over the footprint by featuring Aboriginal musicians and elders who address people throughout it, by emphasizing Aboriginal history and struggles that have transpired across it, and by addressing Aboriginal viewers as driving through it, downlinking within it, and celebrating their control over it.

How might one read the juxtaposition of such Aboriginal shows with syndicated U.S. programs? While *Corroborree Rock* targets an intergenerational audience by bringing elders and youth in a celebration of Aboriginal music and cultural survival, the series that follows, *Dawson's Creek*, does so by interweaving baby-boomer and youth characters in melodramatic plots. Interestingly, both *Dawson's Creek* and *Nganampa* feature communities and stories set near water bodies. But while *Nganampa* emphasizes Aboriginal struggles to regenerate their Fitzroy River water resources, in *Dawson's Creek* water serves as a beautiful backdrop for residents of a fictional white community. What I am getting at here is the idea that we must imagine "flow" and "footprint" not as fixed schedules and closed boundaries but as zones of situated knowledges and cultural incongruities that may compel struggles for cultural survival rather than simply suppress them. In some circumstances the intermingling of Aboriginal and foreign television series in Imparja's footprint and flow may bring structural inequalities into new relief.[57] For instance, we might ask what it means to air "*ER* in the outback?" Might inserting *ER*, and its fictions of immediate and effective health care, in Imparja's footprint call further attention to inadequate medical facilities and treatment for Aboriginal Australians? As Imparja employee Donna Campbell told me in an interview, there are simply not enough Imparja shows that deal with Aboriginal health, edu-

cation, and unemployment.[58] To air a British program like *Neighbours* in
the afternoon, when children are coming home from school, is to raise the
question whether whites and Aboriginal people are living as neighbors and
whether children in the region are being socialized to imagine themselves
as part of a multiethnic community. A program title like *Sale of the Century*
might not just conjure up a game show but may create an uncanny allu-
sion to the forced selling of Aboriginal lands for mining and tourism during the twentieth century. I offer these speculations certainly not to speak
for Aboriginal viewers but to suggest the character and range of contradictions and possibilities that may shape footprints and flows in postcolonial
conditions.

The hybridities generated by Imparja's schedule may also energize Aboriginal revampings of Western television forms. In her discussion of postcolonial literature Benita Parry suggests, "In the 'hybrid moment' what the
native rewrites is not a copy of the colonialist original, but a qualitatively
different thing-in-itself, where misreadings and incongruities expose the
uncertainties and ambivalences of the colonialist text and deny it an autho-
rizing presence. Thus a textual insurrection against the discourse of cultural authority is located in the natives' interrogation of the English book
within the terms of their own system of cultural meanings."[59] Although
Parry is discussing reading and writing in postcolonial contexts, the same
principles can be applied to satellite television in Australia. Aboriginal Australians who own and operate Imparja tv use satellites not just to downlink,
dub, and rerun American and British programs and flow structures; rather,
they select shows and arrange them in ways that give shape and meaning
to the Imparja footprint. What Parry's comment implies is that Imparja is
not simply "copying" or rebroadcasting American or British programs but
that their very selection and arrangement constitutes a different "thing-in-
itself." Imparja's flow can be conceived as hybrid in the sense that it represents a rewriting or reconfiguration of television programming made for
audiences elsewhere.

In some cases Imparja engages in more explicit rewritings of U.S. and
European television forms. For example, as an alternative to shows like
Dallas and *Melrose Place* caama announced plans in the late 1990s to pro-
duce an "outback soapie" entitled *Glen Helen*, which would feature Western

Arrernte landowners and white actors and be set in "a truly unique Outback hotel" called the Glen Helen. Producer Matthew Flanagan developed *Glen Helen* because, as he explains, "*A Town Like Alice* needs to be brought up to date. In this frontier setting, where white and indigenous Australia mix, we also have a crossroads for the entire world."[60] The plan for such a series indicates that the hybridities in Imparja's footprint may inspire fictional television worlds that place multiracial communities together not only in the same flow but in the same show. Unfortunately, when I visited Imparja in 1999, I discovered that funding for *Glen Helen* had fallen through and its production was on hold.

Satellite Downtime/Uptime

In postcolonial conditions the very capacity to produce footprints and flows is a significant form of power, for it represents high levels of institutional organization, technological knowledge, and political coordination. Imparja's flow involves the purchasing and rescheduling of Western programs, synchronizing with Aboriginal celebrations, and conceiving of the signal as carrying and connecting remote Aboriginal communities. Just as important as Imparja's flow are the network's practices of distribution. Imparja uses satellite television in ways that blur distinctions between cultural production and distribution with its imaginative and resourceful regulation of transponder time. Don Browne suggests that Aboriginal broadcasters in urban and remote areas have largely adopted the European and American broadcasting principle of prime time, which airs the most profitable programming during the evening hours.[61] But during the early years of its operations Imparja's broadcasters also used the network's sign-off hours—a period I will refer to as the *satellite downtime*.

Shortly after Imparja won its license in 1986, it developed a resourceful use of the satellite designed to balance the organization's funding shortage with its goal of transmitting as much Aboriginal programming as possible. The plan involved using satellite downtime to distribute dreaming videos to Aboriginal communities in the central zone. When the transponder was not being used to relay its commercial signal, Imparja staff would send videos to local broadcast outlets throughout the footprint. These programs could then be substituted for existing programs in local transmis-

sion schedules after villages discussed the matter and voted.[62] During the
satellite downtime Imparja became a relay center for the circulation of Aboriginal dreaming videos, many of which contained sacred performances
considered unsuitable for outsiders' eyes.[63]

This practice was feasible only because communities such as Yuendumu
in the Northern Territory and Ernabella in South Australia had been producing videos since the early 1980s. Eric Michaels's research revealed that
between 1982 and 1986 the Yuendumu community produced and archived
more than three hundred hours of videotape. In 1983 the Pitjantjatjara
people (a community of 550) formed the Ernabella Video Project and began
producing local videos that reflected the interests and activities of the community.[64] The shows became so popular that they were distributed to different Pitjantjatjara communities. Eventually, crews from Ernabella began
traveling to different communities to produce videos about their people.
While most of these tapes were kept in communities, some of them have
been shared via satellite either during the downtime or more recently on
Nganampa.

To get a sense of the kind of material relayed in the satellite downtime,
we might consider Eric Michaels's discussion of the Yuendumu video *Coniston Story*. This video is a recording of an oral narrative delivered by an elder
called Japangardi, who describes the events he witnessed as a child along
Crown Creek on what is now Coniston Station, seventy kilometers east of
Yuendumu. In 1929 Aborigines axed white trapper and dingo hunter Frederick Brooks to death. Later police arrived at the site and murdered almost
one hundred Warlpiri people to punish them. This period is known as "The
Killing Time," and Michaels suggests that this story "has come to function
like an origin myth, explaining the presence and nature of Europeans and
articulating the relations that arose between the two cultures."[65] To prepare a video account of the event, producers took twenty-eight people to
Coniston to witness the retelling of the story. The piece, Michaels explains,
is filled with long takes of storytelling by Japangardi at various sites associated with the murder. Many long slow pans across the landscape are
said to "follow the movements of unseen characters . . . which converge on
this landscape" as camera operator Jupurrurla explains, " 'This is where the
police trackers came over the hill,' 'that is the direction the ancestors come
in from.' "[66]

Coniston Story had validity as an official account of the story because it was performed and recorded in the presence of many witnesses, particularly tribal elders. Aboriginal societies are extremely conscious of who has the power to tell a story and who has the power to hear a story. As a result, Aboriginal societies have developed strict rules regarding the circulation of dreaming videos. In some cases the tapes cannot be shown to generalized audiences, and they must be destroyed or locked away in an archive if an elder that appears dies. The dreaming videos are not static records of past events; instead, they are imagined as actively forging connections between storyteller, listener-viewer, and physical environment with each retelling and replaying.[67]

Imparja's use of the downtime, then, allowed the network to take advantage of the satellite's ability to extend community ties across its large footprint while still conforming to Aboriginal customs regarding the sanctity of certain representations and stories on video. Within most Aboriginal cultures, access to certain dreaming stories is limited to those within a linguistic group or kinship line and to those of a particular rank in the community hierarchy of elders. Thus, while Imparja was founded in part to circulate and preserve Aboriginal cultural forms, in some cases it might be inappropriate for the network to broadcast dreaming videos widely via satellite. The tactical use of satellite downtime, then, facilitated the regional distribution of Aboriginal tapes that were appropriate for community viewing but not for viewing by white Australians watching the Imparja signal. This use of the satellite as a technology of point-to-point uplinking and downlinking, rather than broadcasting, enabled Aboriginals to craft a distribution practice of microcasting rather than broadcasting. As Hamid Naficy has shown, narrowcast and lowcast forms of television can be "instrumental in helping displaced populations form and maintain cultural identities from a distance and across national and geographic borders."[68] By transmitting literally at the margins of its broader commercial satellite television services, Imparja devised a distribution system that encouraged the production of Aboriginal videos while keeping the stories and knowledge within them close to home.[69]

By the mid-1990s Imparja no longer used the downtime, and the network seized the convergence of satellite, television, and digital technologies to reinforce its global presence, articulating what might be called a *satellite uptime.* In contrast to the satellite downtime, then, the satellite up-time involves inscribing Aboriginal presence within the global media econ-omy. In 1996 Imparja put itself on the World Wide Web with a Web site made of Aboriginal iconography and emphasizing the network's Aborigi-nal ownership. In 1998 Imparja underwent a conversion to digital satellite delivery before any of the other Australian television networks.[70] As a small network it could convert its system more quickly, and in the process Im-parja's audience grew from 200,000 to 430,000.[71] This digital conversion also gave Imparja control of an additional channel, which became the Na-tional Indigenous Television Service in 1999. This, in effect, institutional-ized the downtime, creating a channel exclusively for Aboriginal use. One Aboriginal leader described it as a "cultural exchange of Indigenous infor-mation and entertainment on a level never achieved before."[72]

As Imparja has expanded its operations, recent episodes of *Nganampa* and *Imparja National News* are symbolic of its efforts to balance Aborigi-nals' global presence with local autonomy. A two-part *Nganampa* episode titled "Palm Valley to Lombok" features a group of Aboriginal artists known as the Hermannsburg potters as they journey from the Nataria commu-nity to Lombok, Indonesia. The episode emphasizes the women's move-ment abroad: We see them traveling away from their community by car, riding on a boat, getting on a Qantas airplane, hopping on a bus in Lom-bok, and walking through villages where they meet the Sasale people, who have been making ceramic pots for hundreds of years. While in Lombok the women learn pottery techniques, get massages, chew tobacco, and show their paintings, but their travels ultimately reinforce their desire to re-turn home. As one of them explains, "I felt so lonely in that country I started to cry." Another says, "I didn't like those noisy people looking hard at us. . . . I was so ashamed." The episode acknowledges the Aboriginal artists' ambivalence about global presence and closes in their home terri-tory of Palm Valley, where they retell the dreaming stories of their region. Another two-part *Nganampa* episode, titled "The Life of Ginger Riley," fea-tures the internationally renowned Aboriginal artist as he moves between his mother's home in the Limmen Bright region of southeast Arnhem Land to national and international art galleries. The episode emphasizes his Ab-original roots and global fame, mentioning an international prize in 1992, his commission to create pieces for the Australian embassy in Beijing, the stamp he designed in 1994 to honor the International Year of the World's Indigenous Peoples, and recent exhibits of his work in London and Japan. As Riley stresses to viewers, "The white man didn't tell me anything about painting."

While Imparja used the satellite downtime to maintain control over sacred Aboriginal videos, its circulation of these (and other) *Nganampa* episodes places them within a global television lineup where they are screened alongside television from abroad. By emphasizing the international migration of Aboriginal artists and connecting them to Aboriginal lands, the episodes are symptomatic of Imparja's own efforts to move Aboriginal culture up to a satellite while constantly returning home to serve the interests of Aboriginal people. One of the most important Imparja programs in this regard is the nightly news program, anchored, produced, and shot by Aboriginal people who have either been trained at Imparja or at national media training centers. While Imparja filters almost all of its fictional programs, much of its news coverage comes from Aboriginal people. This is particularly significant given that Aboriginals are often overlooked or stereotyped by TV news agencies in large Australian cities such as Sydney, Perth, and Darwin. Imparja employees take great pride in their nightly news program, which opens with Imparja's logo and the sounds of a didgeridoo and then combines feeds from ABC, SBS, and CNN with locally produced segments.[73] Here again Aboriginal television is positioned within a global media economy. For example, in July 1999 international coverage of the war in Kosovo, political demonstrations in Iran, and civil turmoil in Indonesia and East Timor were followed by reports on the Yuendumu tribal elders' visit with the Australian national labor chief, the Warlpiri Media Association's release of a new Yothu Yindi CD, and the opening of an Aboriginal health center. By mixing global, national, and local signals *Imparja National News* encourages viewers to imagine their communities in relation to global events—and does not construct the world as a harmonious global village but often as a sphere of social and political strife.

On some occasions Imparja news stories have been "cherry picked" by other news organizations worldwide. In 1999, for example, Imparja TV camera operator Dwayne Tickner worked on a story called "Stolen Generations," about Aboriginal children who had been taken away from their parents as infants and placed in white homes during the 1950s, videotaping members of this generation as they reunited with their Aboriginal parents.

This would have been a typical local Imparja segment, but as Dwayne explained, all the Australian networks and CNN jumped on it, and his footage was relayed via satellite around the world.[74] The global circulation of this news segment was particularly significant since it exposed what had happened to a generation of Aboriginal babies delivered into the hands of white colonizers. The story was followed up in Australian national news reports and even became part of the television coverage of the Sydney Olympics in 2000 and the internationally distributed feature film *Rabbit-Proof Fence* (2002). Like the outward moves of the Hermannsburg potters and Ginger Riley, the "Stolen Generations" segment represented a global migration for Dwayne Tickner and Imparja TV.

These practices of the satellite downtime and uptime suggest that we should imagine television not only as a form of horizontal fragmentation (or flow) but also as a set of *vertical* practices of uplinking and downlinking. Gilles Deleuze uses the term *verticality* to critique the hierarchical logic embedded in Western modernity, but I invoke it to refer to television's multidimensionality and territorializing properties, particularly as it converges with satellite technologies. Television is not just what appears on-screen; it is a variety of invisible yet specific practices that occur in the air, in orbit, and across lands. Geoffrey Batchen goes so far as to suggest that we could reconceptualize television as "an indiscriminate and all encompassing atmosphere of electronic data, a field of impulses that continually surrounds and traverses us whether a monitor is present or not."[75] For decades television signals have moved through the air and orbit, bouncing from one geographic territory to another, but we have tended to conceptualize and analyze these distribution processes as distinct from the material on the screen. *Our World* attempted to visualize these vertical practices in dramatic and spectacular ways, but they have since become invisible, naturalized as global television's seamless operation and ceaseless flows. If, in the context of television's globalization, flow and footprints are generated through the migration and filtering of signals from some parts of the world to others, then the power to access, regulate, and contribute to these vertical modes of exchange is more significant than ever.

What is perhaps unique about Imparja is that its strong emphasis on Aboriginal lands brings the relationship between satellite television and territory to the fore. As Eric Michaels has explained, Aboriginal interest in "air-rights" has always been tied to other components of Aboriginal self-determination, such as land rights, the outstation movement, and uses of technologies such as cars.[76] In Western societies the broadcast frequency spectrum is treated as a natural resource or property and is either licensed to "public service" broadcasters or sold off to corporate entities. In Aboriginal Australia, satellite television is a commercial operation, but the technologies have been imagined as an extension of the landscape. This perception is exemplified in Imparja's station identification spots that combine Aboriginal dreaming paintings, digital animation, and realist video perspectives of rock formations, wildlife, and rivers throughout the footprint. It is emphasized in episodes of *Nganampa*, which are relayed via satellite to "air" the dreaming electronically, and in Aboriginal music videos, which are purposely performed and shot "in the bush." It is perhaps most explicitly suggested in the term *satellite dreaming*, which I discuss in the next section. What we can learn from Imparja's downtime and uptime is that struggles over what Eric Michaels called "the cultural future of television" exist not only within the studio or the frame but also in the ways signals move through the air, orbit, and across territories. Imparja TV is useful to television criticism precisely because its practices provoke further deliberation of the relationship of footprints and flows to geophysical locations, spatial imaginaries, and territorial practices. Consider, for instance, that Imparja's footprint is defined not only by the topography of Australia's central desert but also by such imaginary elsewheres as *Dallas*, *Melrose Place*, and *Dawson's Creek*. Just as flow becomes a site of hybridity in the age of satellite filtering, the satellite footprint becomes a cultural topography in which spatial and temporal imaginaries accumulate, mix, sediment, and stir. The emergence of satellite television networks has triggered so much anxiety about national sovereignty and cultural contamination over the past two decades precisely because footprints represent the power to transform, redefine, and hybridize nations, territories, and cultures in a most material way.

Satellite Dreaming

There are two Aboriginal artworks called *Satellite Dreaming*. One is an internationally distributed CAAMA documentary about Aboriginal television in Central Australia and the Northern Territory. The other is a locally exhib-

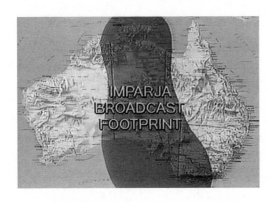

The first Imparja footprint
presented in the documentary
Satellite Dreaming.

ited acrylic painting by Andrew Japaljarri Spencer. The fact that Aboriginals have incorporated "satellite dreaming" into their narratives of cultural
identity is highly symbolic and important. Since dreaming stories are typically identified in English by the name of the creature that moved through
and contoured the landscape (for example, "Caterpillar Dreaming," "Emu
Dreaming," and so forth), this gesture posits the satellite itself as a creature of the dreaming—as part of Aboriginal cosmology. It also suggests that
the satellite has become part of Aboriginal storytelling practices. "Satellite
dreaming" thus alludes to ways in which the satellite—as a technology of
Aboriginal territorial reclamation and cultural survival—has altered both
the landscape of Central Australia and the Aboriginal cultures that thrive
on it.

The CAAMA documentary *Satellite Dreaming* explores Aboriginal television practices in Alice Springs, Yuendumu, and Ernabella. Released in
1991, the video recounts Aboriginals' battle for the central zone footprint
in the mid-1980s and explores the significance of television for Aboriginals
by profiling Aboriginal and non-Aboriginal media practitioners from outback and urban communities. *Satellite Dreaming* opens by establishing a
connection between the orbiting satellite and the Aboriginal videographer.
The segment begins with a rocket launch as a bellow of flames propels
the rocket into space. A small communications satellite emerges from the
rocket capsule and enters its orbital path. As a transition to live action, the
image dissolves from the deep indigo of outer space to the cerulean sky of
the outback, followed by a tilt down to an auburn dirt road. As the whirring
sounds of outer space fade, we hear chanting voices, and we see Aboriginals

driving through the bush. The Pitjantjatjara people of Ernabella, we learn, have been invited to videotape the Seven Sisters dreaming in Kuruala.

The sequence structures an interesting comparison between the satellite in outer space and Aboriginals in the outback. Editing juxtaposes images of a satellite floating in orbit with images of Aboriginals traveling along dirt roads, exploring their similar patterns of movement. And while the satellite relays audiovisual signals, so, too, does the Pitjantjatjara video crew. There is, in other words, an important connection established here between local Aboriginal cultural production and satellite distribution. Aboriginals have historically reenacted their relationship to the land by traveling widely within it, and here video and satellite technologies become extensions of this same process. Moreover, the segment suggests a similarity in the spatial position of Aboriginal cultures and that of the satellite, for remote Aboriginal lands might be considered satellite territories in the sense that they are mapped around centers of Australian settlers. Further, individual Aboriginal communities themselves are satellitized—that is, they are spread across vast distances, linked together by the cultural apparatuses of the dreaming lines. The iconography of the segment, however, links the satellite in outer space to the Aboriginal communities of Central Australia. In a sense, then, it both maps Aboriginal territory into orbit and incorporates the satellite within Aboriginal discourses on the landscape itself.

An acrylic painting by Andrew Japaljarri Spencer, also titled *Satellite Dreaming*, forges a similar connection between the satellite and Aboriginality. The painting serves as the logo for Yuendumu broadcasts. According to anthropologist Robert Hodge, it "shows a set of concentric circles at the centre, representing Yuendumu, with 16 tracks radiating out to 16 other sites, of different sizes, all consisting of concentric circles. This uses classic Warlpiri iconography to represent Yuendumu as the complex centre of the social community and the media universe."[77] The painting also appears in CAAMA's *Satellite Dreaming* documentary, where it is featured in a segment on a 1990 satellite teleconference between two Aboriginal communities. A member of the Yuendumu community points to the painting, which he uses as a cultural map to demonstrate how live outstation video connections will be possible with new satellite technology.

These *Satellite Dreaming*s symbolize the integration of satellite technology within the cultural traditions (and especially the storytelling prac-

Satellite dishes are scattered throughout remote parts of Australia. Photo by the author.

tices) of Central Australian Aboriginal communities. Both the documentary and the painting construct satellite technology as part of Aboriginal cultural production. Rather than destroying local cultures, CAAMA executive Philip Batty claims, the satellite has become part of an Aboriginal "cultural renaissance."[78] The integration of satellite television within Aboriginal communities has led to other kinds of satellite dreamings as well, such as Aboriginal artist Tracey Spencer's painting of dreaming tracks on *Landsat 7*'s remote-sensing satellite images of the Northern Territory.[79]

Such Aboriginal cultural discourses are particularly significant in the wake of earlier uses of satellite television, which imagined the technology as beyond the reach of non-Western societies or used it in an effort to "develop" them. Aboriginal satellite dreamings also challenge critical assumptions that satellite television works *only* as an agent of Western cultural imperialism and neocolonial control. Such an interpretation is based on a romantic, even fetishistic, reverence for indigenous cultures as detached relics of an ancient past. As Batty puts it, "It would be a comfort perhaps to romantics who tend to regard Aboriginal culture as a static entity, immune from change, located forever in some timeless never-never land, to believe that Aboriginals have rejected the 'evils' of global television and banned the

intrusion of such spiritually impure rubbish from their communities. Most
people, including Aboriginal people, aren't like that. Aboriginal kids love
Kung Fu movies and Michael Jackson clips even if they can't understand
the lyrics."[80] Some refuse to recognize the impossibility of a "pure" Aboriginal culture and tend to regard it as something that should be insulated and
protected from external influences. We should complicate paternalistic arguments that pit "cultural preservation" against "Western corruption" and
look instead at how satellite television has been used by Aboriginal owners
to negotiate spheres of global, national, and local cultural and economic
activity and to fuel ongoing struggles for cultural survival and territorial
reclamation.

Conclusion

Although Imparja's circulation of Aboriginal programming is limited, this
satellite television network serves vital cultural functions and has significant political effects. On the one hand, Imparja represents an Aboriginal
effort to engage in a struggle over satellite technology at a broad structural
level—to inscribe an Aboriginal presence across the broad footprint of *Aussat*'s path and beyond it. On the other hand, the contingent and situated
nature of Aboriginal knowledges makes it impossible for the satellite network to become a unified pan-Aboriginal medium. As a result, small-scale
Aboriginal television production and distribution systems have emerged,
circulating highly localized texts whose meanings might be at odds not only
with white Australian and American television but with the pan-Aboriginal
Imparja as well. Aboriginal Australians have asserted control within both
physical and mediated spaces, revived and relayed dreaming stories, and
responded to the dominance of Western television flows.

Throughout this chapter I have suggested that Imparja TV serves as a
valuable site for rethinking how the concepts of the satellite footprint and
television flow might be understood in the context of postcolonial conditions. Imparja's flow functions as a site of cultural hybridity, exposing
the ambivalences and continuities between Aboriginal and Western television cultures. The Imparja footprint is not only a boundary in which the
network's signal falls; it is also a cultural topography in which Aboriginal
dreamings and Western televisual fictions collide. Through it Australian

Aboriginals have staked a claim to a new electronic space that overlaps with their physical territory and their cultural autonomy. Despite tendencies in the 1960s to use satellite television to assert the primacy of Western industrial nations, Imparja TV demonstrates the possibility of using the technology to support the autonomy of social minorities, although never in a pure or total way. As Stanley Aronowitz reminds us, "It is crucial to avoid positing universal descriptions or analyses of what it is that technologies mean . . . and do (both in terms of their hegemonic power and functional capabilities)."[81] In this chapter I have described satellite television as a technology of Aboriginal cultural survival and territorial reclamation. The discussion of postcolonial minorities' relationships to satellite footprints and television flow is particularly important during a time in which Rupert Murdoch's Newscorp aspires to create satellite television services that span the entire planet (presumably to also "serve" remote communities). But unlike Imparja, Newscorp lacks a mandate to serve the interests of indigenous peoples or any people, for that matter. Imparja can thus be understood as an important counterpoint to multinational media conglomerates like Newscorp; although Imparja, too, operates within a global media economy, it uses its footprint and flow in an effort to represent the interests of postcolonial minorities. Rather than assume, as Western critics have historically done, that developing nations, indigenous peoples, or postcolonial immigrants would be ill-equipped to manage satellite technologies, we should assume that subordinated social formations might use satellite television (whether an entire network or simply a satellite dish) to negotiate the fallout of globalization. In *Hybrid Cultures* Nestor Garcia Canclini reminds us that "communications technologies and the industrial reorganization of culture do not replace traditions, nor homogenously massify them, but rather change the conditions for obtaining and renewing knowledge and sensitivity."[82] Since satellite television services now operate on every continent, there is a need for television critics to explore how these technologies are changing conditions for obtaining and renewing knowledges and sensitivities across the planet.

The village of Srebrenica from the perspective of a hilltop overrun by Bosnian Serb troops. Photo by the author.

Satellite Witnessing

VIEWS AND COVERAGE OF THE WAR IN BOSNIA

When technologies converge, they develop in discursive, economic, and institutional interdependence with one another. Convergence, then, is a relational model of understanding how technologies inflect, inform, and interact with one another in processes of their emergence. A genealogy of convergence does not celebrate the newness of combined technological forms so much as it emphasizes the paths of contradiction and ambivalence elicited by their mutual interactions. Treating satellite television as a site of convergence thus involves exploring its disparate forms and practices alongside and in relation to one another. Only by perceiving it in this way can we understand satellite television as a set of discursive modalities that can be articulated across different hardware systems, sociohistorical conditions, institutional settings, and instances of use.

To draw out the complex trails of convergence, the next two chapters shift away from conventional definitions of satellite television as a form of live transmission or broadcasting to examine more militaristic and scientific configurations. Specifically, I consider satellite television as an institutionalized practice of remote sensing—a practice of Earth imaging that, as Jody Berland explains, arises "from the complex imperatives and alliances of three interdependent industries: paramilitary space exploration; computer software; television."[1] Rather than focus on the overlapping industrial origins of remote sensing, I stress its aesthetic-phenomenological components, exploring how its use "affects our experience in ways that are not bound to questions of function."[2] I define *remote sensing* as a televisual practice that has been articulated with military and scientific uses of satellites to monitor, historicize, and visualize events on Earth.

Top-secret military spy satellites generate high-resolution image intelligence that is typically differentiated from the lower resolution practices of "remote sensing." But rather than getting entangled in this technical distinction, I use the term *remote sensing* generally to refer to a practice of orbital viewing that simultaneously extends and exposes particular regimes of vision and knowledge. More specifically, I examine the remote-sensing practices of U.S. military officials monitoring the war in Bosnia and of French archaeologists excavating ancient ruins in Egypt. By treating military and archaeological uses of remote sensing as sites of television analysis, I hope to highlight the multiple ways of constituting satellite television and to complicate the dominant tendency to equate orbital vision with the distant observer of the Enlightenment.

A Brief History of Remote Sensing

Designed as technologies of espionage, remote-sensing satellites were first launched by the United States during the early 1960s as part of the top-secret *Corona* program.[3] *Corona* satellites, which were hidden inside of the high-profile *Discoverer* spacecraft, secretly monitored the Soviet Union, Eastern Europe, and Asia from 1960 to 1972, showing nuclear weapons facilities, long-range airfields, missile manufacturing plants, and lunar launch pads.[4] In 1972, the year the *Corona* project ended, NASA launched the first public remote-sensing satellite, *Landsat*, which took unclassified pictures of the earth for use in natural resource management. Satellite image requests came from government agencies, universities, industry, foreign governments, and high school students who wanted pictures of their hometowns.[5] *Landsat 1* transmitted more than three hundred thousand images of the earth before it retired in 1978 and was followed by a series of other satellites whose images were used in different ways.[6] For instance, *Landsat 5* observed Chernobyl's nuclear meltdown in 1986, and in 1993 it exposed severe deforestation in the Brazilian Amazon. The *Landsat* program was celebrated as fostering natural resource development, environmental observation, and cold war détente. Combined, however, *Corona* and *Landsat* represented military and scientific control of remote sensing, which, I will argue, impacted certain later configurations of satellite television as well. Remote sensing extended tele-vision into orbit and equipped military officials and scientists with a unique vantage point that authorized

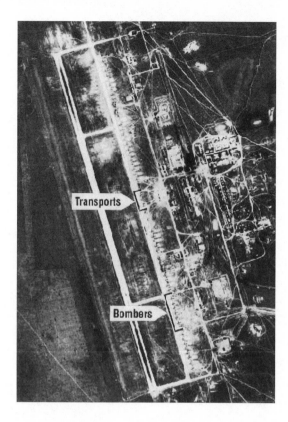

Transports

Bombers

Declassified *Corona* satellite image of a Soviet long-range aviation airfield taken August 20, 1966.

their knowledge practices of intelligence gathering and Earth observation. As satellite spectaculars such as *Our World* celebrated the satellite's potential to bring the world together in a global now, they concealed the fact that in that same instant satellites were encircling the earth on planetary patrols. But just as *Our World* reinforced the global presence of the Western viewer, military and scientific remote-sensing programs treated the surface of the earth as a domain of unobstructed Western vision, knowledge, and control. As space analyst John McElroy has suggested, "Observations of the earth from space know no national boundaries. The passage of a satellite from the space above one country to that above another requires no visa: technology has made obsolete the concept of complete national privacy."[7]

During the 1980s satellite trespassing became even more pronounced as a host of industrial nations and international companies deployed remote-sensing satellites that competed with the *Landsat* series. In the mid-1980s the French company that owned SPOT (*Satellite Pour l'Observation de la Terra*) began selling high-resolution images in the global market, and in 1987 the Soviet company Soyuzkarta joined the enterprise as well. In 1994 President Clinton privatized the U.S. remote-sensing industry and invited American companies to apply for licenses that would allow them to sell satellite images of up to one meter in resolution. This policy was significant since for decades only military intelligence analysts had regular access to such images. In the mid-1990s U.S. companies such as Earthwatch and Space Imaging emerged, promising high-resolution images to any individual, nation, or corporation with the financial resources to purchase them. Since then the satellite image—once a top-secret intelligence medium—has become increasingly available to the public. As the *New York Times* proclaimed in 1997, "Commercial spy satellites are about to let anyone with a credit card peer down from the heavens into the compounds of dictators or the backyards of neighbors with high fences."[8] Satellite images are now used by real estate agents, archaeologists, city planners, refugee relief agencies, meteorologists, criminal prosecutors, cartographers, travel agents, television news producers, and science teachers, to name but a few.[9] They have been used to find the lost ancient city of Ubar, track the movement of Rwandan refugees, monitor the ebbs and flows of El Niño, count the participants in the Million Man March, and search for O. J. Simpson's white Ford Bronco.

As the uses of satellite images expand, they have also been integrated within television news coverage. In 1989 ABC was the first U.S. television network to broadcast a strategic satellite image when it aired a *SPOT* photo of a Libyan chemical weapons facility.[10] In the late 1980s such satellite images took months to process and analyze. Since then, however, digitization has expedited satellite image processing dramatically. The president of a remote-sensing company called Eyeglass promises, "We [will] . . . be able to downlink real-time imagery of any point on the earth. . . . Our system will have the ability to pass over every point in two days or less."[11] This means that a satellite image of a newsworthy event such as a flood, earthquake, or nuclear accident can be "passed on to television networks within hours, not days." In 1994 such images were projected to cost about two thousand dollars.[12] In 2001 that figure dropped to three hundred dollars. Satellite images have also become standard in television news coverage of global events, such as the war in former Yugoslavia, nuclear weapons tests in India and Pakistan, the 9/11 attacks on the Pentagon and World Trade Center, and the wars in Afghanistan and Iraq.

Because of their long association with military and scientific institutions, satellite images are often framed as the most "official" and "objective" view of an event. In order to complicate this assumption, the next two chapters critically examine the production and circulation of military and scientific satellite images and their integration within mass cultural discourses. Together these chapters challenge the correspondences forged all too often between distant observation, objectivity, and truth. I suggest different ways of engaging with remote-sensing images, treated here as different forms of satellite television, imagining them as possible sites of witnessing and excavation, that is, as empty fields of signification that necessitate closer readings, semiotic infusion, embodiment, and struggles over interpretation.

Satellite Oversight

The images on the following pages were acquired by U.S. military satellites in July 1995. They reveal alleged mass-grave sites in eastern Bosnia, and their public release by the U.S. State Department in August 1995 initiated a United Nations investigation into what is now known as the Srebrenica massacre. Filmmakers, journalists, military officials, human rights organizations, and even forensic anthropologists have tried to document the events that took place in Srebrenica in July 1995.[13] Much of the media discourse surrounding the event has placed the blame solely on the Bosnian Serb Army (BSA) officers who ordered and oversaw the takeover of the UN-protected safe haven. To make complex events more comprehensible, the Western media tended to demonize the Serbs when covering the war in Yugoslavia.[14] Recent reports, however, reveal that the Bosnian Muslim Army also committed atrocities near Srebrenica (in Bratunac, for example) during the war and must take partial responsibility for conditions leading to the massacre.[15] Since 1995 UN prosecutors have indicted thirty-eight suspected Bosnian Serb and Muslim war criminals to testify before the War Crimes Tribunal, and many of these trials are still underway as of this writing.[16] What has not been examined in such detail, however, is the United States' satellite oversight of these events.

Declassified U.S. satellite image acquired in July 1995, showing possible mass graves near Srebrenica.

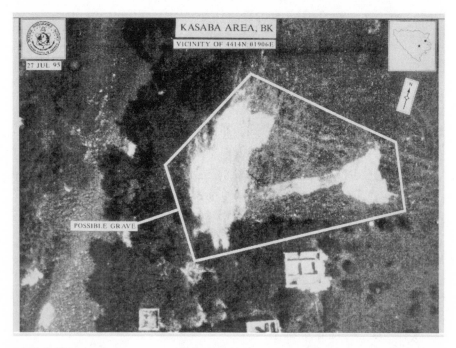

Declassified U.S. satellite image acquired July 27, 1995, showing a possible mass grave near Srebrenica.

Journalists and human rights organizations have described the events
that took place in Srebrenica from July 6 to July 17, 1995, as "the worst
atrocities of the Balkans war" and as "Europe's worst massacre since World
War II."[17] During this two-week period thousands of Muslim men disap-
peared from the UN-protected safe haven as U.S. satellites passed overhead
to monitor the region. American intelligence agencies supporting the U.S.-
NATO peacekeeping mission in Bosnia acquired satellite image data related
to the Bosnian Serb Army's takeover of Srebrenica but failed to release it
in a timely manner. In the age of "virtual warfare" and in the context of the
U.S. military's campaign of Information Dominance, the delayed circulation of satellite intelligence could itself be considered an atrocity, particularly when the activities in view are deemed "genocidal."[18]

This chapter critically examines American broadcast news segments
containing U.S. satellite images of mass graves and considers how state
satellite intelligence became part of the television coverage of the war in
Bosnia. The events that occurred in Srebrenica in July 1995 could never be
reduced to or contained within the satellite images that ostensibly captured
and revealed mass executions unfolding on the ground below. Despite their
claims to "objectivity" and "omniscience," these images represent several
satellite "passes" over Srebrenica at a particular time on a particular day,
and in this sense they are extremely partial and selective. Nevertheless they
do represent an attempt to regulate the meanings of the war from orbit. U.S.
officials claim that their satellites acquired evidence of genocide, but these
images really only scratch the surface, so to speak, of events that unfolded
on the ground below. In this sense the satellite images of mass graves in
Bosnia are highly symptomatic of the U.S. position vis-à-vis the Balkans
more generally—that is, its distant and technologized monitoring, and its
refusal to acknowledge (put into discourse) the complex political, sociohistorical, economic, and cultural conditions that gave rise to recent conflicts
in the former Yugoslavia.

In an age of technologized vision one of the most important functions
of the witness is to demilitarize military perspectives—that is, to open the
satellite image (and other forms of image data and intelligence) to a range
of critical practices and uses. To address these issues I explore what it might
mean to "witness" wartime atrocities from the perspective of an orbiting
satellite, and I use the case of the Srebrenica massacre to demonstrate the
need for greater public knowledge of satellite technologies and increased

literacy around the media they generate. The satellite image's aesthetics of remoteness and abstraction make its status as a document of truth very uncertain and unstable and open it to a range of possible interpretations and political uses. In this case the satellite image functions not only as the state's official perspective but also to implicate the state in whatever lies in the field of vision. In addition to specifying the location of alleged mass graves, satellite images of Srebrenica spotlight the passive-aggressive voyeurism of U.S. Information Dominance and the politics of ethnic cleansing, a politics in which the United States publicly condemned Bosnian Serb Army aggression while idly recording its attack on Srebrenica's Muslims.

Event

To recount what happened on the ground in Srebrenica in July 1995 is an almost impossible task. My own description of these events is based on my distance from them, on my attempts to understand them through U.S. media and English translations emerging from former Yugoslavia. What follows is not an attempt to provide "the truthful account" of the Srebrenica massacre but rather an attempt to approach it as a media event, an event reimagined based on and filtered through various press reports and media coverage. John Fiske suggests that "media events are sites of maximum visibility and maximum turbulence," that they are "spectacles that people watch from the safe distance of specially constructed viewing platforms."[19] The fact that I wrote the following description from a "safe distance" means that it is laden with particular assumptions about who was responsible for the Srebrenica massacre, assumptions that I ultimately want to confront, complicate, and unravel. Nevertheless, I must start somewhere, so it is here, as a concerned but skeptical media scholar, that I begin.

On June 17, 1995, U.S. intelligence officials began tapping telephone lines between Serbia and Bosnian Serb territory, listening almost daily to conversations between Serbian Army chief of staff General Perisic and Bosnian Serb military leader General Ratko Mladic. During those conversations the generals planned the military takeover of the UN-protected "safe haven" in Srebrenica. A few weeks later U.S. aerial intelligence analysts noticed that Bosnian Serb armed forces artillery and tank units had moved to the outskirts of Srebrenica. American military officials chose not to prioritize Srebrenica as an "area of concern," however, for they believed that

Bosnian Muslims were held by Bosnian Serb Army troops at this battery factory in Potočari, and some were allegedly trucked away, killed, and buried in nearby fields. Photo by the author.

the Bosnian Serb armed forces were at most attempting to demilitarize or "neutralize" the enclave.[20]

From July 6 through July 10 the Bosnian Serb Army staged a full take-over of Srebrenica, and this temporary home of more than forty thousand displaced people became a zone of fear and violence. The attack began on July 6, with the firing of rockets into the UN compound, and it lasted nearly two weeks. Bosnian president Alija Izetbegovic warned Sarajevo radio listeners on July 8 that the people of Srebrenica faced "massacres and genocide" if the town was overrun.[21] On July 10 a UN officer typed this desperate plea on his computer: "Urgent, urgent urgent. BSA is entering the town of Srebrenica. Will someone stop this immediately and save these people. Thousands of them are gathering around the hospital please help." And by July 12 a German nurse, working for Doctors without Borders, issued a radio report stating, "Everybody who could have stopped this mass exodus should be forced to feel the panic and desperation of the people."[22]

On July 13 the Bosnian Serb Army followed Dutch peacekeepers and thousands of fleeing Muslims to a hideout in a defunct factory in nearby Potocari. Bosnian Serb forces stormed the enclave, separated "military-

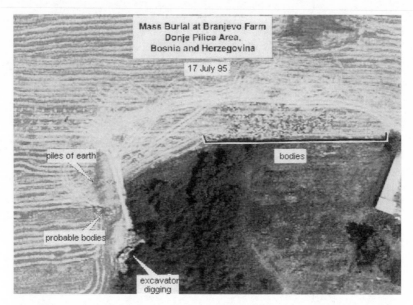

Mass Burial at Branjevo Farm
Donje Pilica Area,
Bosnia and Herzegovina

17 July 95

piles of earth

bodies

probable bodies

excavator
digging

Declassified U.S. satellite image acquired July 17, 1995, showing a mass burial site near Srebrenica.

aged" Muslim men, and transported them in groups of twenty or thirty to nearby fields and abandoned buildings. United Nations soldiers recounted seeing civilians shot in the head, watching them being beaten with rifles, and hearing constant screaming. Two refugees described the body of a man hung from hooks inside the Potocari factory. Others saw civilian Muslim men lying dead on the roadside with sliced throats.[23] The most extensive violence, however, occurred from July 14 to July 17, when Serbian soldiers, apparently wearing stolen UN uniforms and using stolen UN equipment, systematically executed several thousand Muslim men on a soccer field and then bulldozed their bodies into mass graves around the area.

As these events occurred on the ground U.S. intelligence officers used satellites to monitor them from afar. These satellite images of Srebrenica were not circulated publicly until weeks after the massacre occurred, however. On August 10, 1995, the U.S. ambassador to the United Nations, Madeline Albright, presented several of them to the UN Security Council, demonstrating the location of suspected mass graves near Srebrenica. Albright reportedly presented "before" photos of an area in Nova Kasaba, which showed living prisoners crowded into a soccer field, and "after" photos captured a few days later that revealed three areas of disturbed earth

that appeared to be mass graves. She claimed that as many as twenty-seven
hundred Bosnian Muslims had been killed and buried in shallow pits. During her presentation Albright accused General Mladic of "extraordinary
cruelty" and insisted that the satellite images, combined with eyewitness
testimony, provided "compelling evidence that the Bosnian Serbs had systematically executed people who were defenseless with the direct involvement of high-level Bosnian Serb officials."[24] She continued: "The perpe-
trators of these atrocities have literally not covered their tracks. . . . The
physical evidence of what they have done—the bodies discarded in their
fields—will bear silent witness."[25]

Since 1995 many conflicting accounts have emerged about the Srebrenica massacre. Reports from journalists and human rights organizations now speculate that many of the missing Muslim men may have been
killed in battles with Serb forces as they fled the area.[26] Some have suggested that Islamic fundamentalists in the Bosnian Muslim Army provoked the massacre as an act of martyrdom and then killed themselves to
avoid being killed by the BSA. Other reports highlight atrocities committed
by the Bosnian Muslim Army against Serb civilians in Srebrenica from
1992 to 1995 (and during World War II) as a way of explaining how these
events could have erupted. One group even queries, "Was the Srebrenica
Massacre a hoax?"[27] There have also been more general critiques of the
Western media's misinterpretation and sensationalist exploitation of the
term *ethnic cleansing*.[28] The number of dead continues to be a controversial issue. Recent reports from the International Red Cross indicate that
the remains of forty-three hundred victims of the 1995 Srebrenica massacre have been exhumed, but only 118 bodies have been identified.[29] Muslim women from Srebrenica (most of whom now live in Tuzla, Sarajevo, or
elsewhere) are still missing sons, husbands, and brothers, and they have
erected a memorial near the factory in Potocari.[30] I mention these issues
only to highlight the impossibility of knowing what exactly happened in
Srebrenica in July 1995 and to come to terms with my own imperfect attempt to make sense of the events, relying perhaps too readily on quick-fire
journalism and media sound bites. We know that something horrific happened there, but each political interest puts its own spin on the event in
order to extract a strategic maximum from it. Again, this chapter is not a
search for the truth of the Srebrenica massacre.[31] It is an examination of
the shifting coverage of war in an age of technological convergence and

military-media collaboration, and a consideration of how we, as Western citizens and viewers, come to see, know, and comprehend events such as the Srebrenica massacre from afar.

Coverage

In an information society it is almost impossible to differentiate one's knowledge of an event from the media's coverage of it. According to *Bosnia by Television* the war in the former Yugoslavia "was the most recorded and reported of all [military] conflicts,"[32] covered by journalists in the field and satellites in the sky. The term *coverage* is typically used to refer to the stories and images that are generated by journalists and broadcast by networks, but I want to consider it as a particular kind of televisual practice, one that structures the way viewers see and know the world from a distance.[33] The meanings of *coverage* can be specified technologically (as a system of satellite-TV-computer convergence), institutionally (as a set of relations between the state, the media, and citizen-viewers), and textually (as a sequencing of signals and views).[34] At the most basic level, to cover is to record, arrange, and transmit electronic or digital signals—to bring an event to the screen.[35] Coverage requires the use of remote television cameras, which, as Sam Weber reminds us, are mobilized to "see for us."[36] Like a remote television camera, the remote-sensing satellite "sees for us." In a sense its orbital platform establishes a ceiling or vertical limit for televisual representation.

In August 1995 coverage of Madeline Albright's UN presentation appeared on CBS and ABC evening news broadcasts in the United States.[37] This was significant not only because the U.S. State Department rarely releases satellite intelligence in a public forum but also because in this instance the satellite image became part of television's coverage of the war in Bosnia. Consistent with the spatial relations of global presence discussed in chapter 1, the news segments first featured color graphic maps of Bosnia and Srebrenica to establish the geographic location of the event. As news anchors recap Albright's UN presentation, the State Department's black-and-white satellite images burst from points on the map and fill the frame. The words "possible mass graves" and "disturbed earth" were superimposed over the images, and the anchors referred to the images as the state's "incriminating evidence" revealing large plots of recently turned earth in

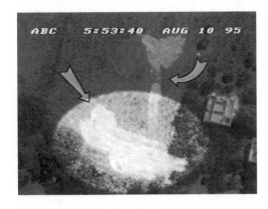

Television news reports integrated satellite images of mass graves in Bosnia but had to rely on shading and arrows to make the images comprehensible to viewers.

Kasaba/Konjevic Polje, near Srebrenica. Each of the mass graves was said to be large enough to hold six hundred bodies, and there were reportedly at least twenty-seven hundred Muslim men missing.

Although the TV anchors described the satellite images as "evidence," there was nothing evident about them. News producers went to great lengths to decode and interpret these satellite images of Bosnia for viewers. Narration and graphics are applied to the satellite images to make them comprehensible and to embed them within a broader economy of televisual signs.[38] In each segment producers used zooms in and out, arrows, and shadowing devices to pinpoint the location of suspected mass graves and to orient the viewer within a classified military perspective that is typically unavailable to citizens' eyes. Since the satellite occupies a position that no human eye can occupy, its views appear to emanate from an unearthly position. As Weber reminds us in his description of television, "What we see (on the television screen) is someone or something seeing. But that someone or something remains at an irreducible, indeterminable distance from the television viewer; and this distance splits the sameness of the instant of perception as well as identity of the place in which such viewing seems to occur."[39] When the viewer encounters satellite images of mass graves on the television screen, then, there is a striking incommensurability between televisual and orbital views. On TV the satellite image of mass graves functions as a *tromp de l'oeil*.[40] Not only does it expose a U.S. military gaze that has for decades been kept strategically invisible, but the intimate medium of television (renowned for its use of the close-up) simply cannot accommodate the excessive abstraction and emptiness of the satellite perspective.[41]

Satellite images privilege the panoramic and the territorial over that of the close-up and the bodily. As much as the narrator tries to anchor these images within television's economy of signs, they evade and disrupt television's conventions, appearing to be quite literally in their own orbit, representing views of the earth that only satellites can see. The gaze of the orbiting satellite exists both within and beyond the system of televisual representation. It can function as one of television's remote cameras (linking the viewer to orbital space), but it also sees the earth from a position that no human camera operator occupies. The gap between satellite and television views is irreducible and disrupts the coherence of the time and place of perception. Hence, the heavy use of graphic devices and narration that works to ground the orbital gaze within the televisual, while also pointing to the highly uncertain status of satellite intelligence as a mass-circulated text.

Since the news segments failed to mention any of the historical conditions leading to the Srebrenica massacre, the satellite images of mass graves function diagnostically as if they were satellite weather photos pinpointing and visualizing the "turbulence" and "chaos" in Bosnia. The televisual coverage is imbued with the Western imperialist notion that the war in former Yugoslavia can be attributed to the "volatile atmospherics of the Balkans" and, as such, is ultimately incomprehensible and unfathomable to those in the "civilized West."[42] The orbital gaze thus promulgates new categories of otherness as it spotlights the world's "trouble spots." As Aida Hozic suggests, "The 'myth of ethnic violence' in Bosnia or Rwanda has helped construe them as unconquerable, ungovernable, even repulsive. . . . The portrayal of these troubled spots as potential quagmires has justified the need for their containment."[43] The U.S. State Department is able to use satellite images of mass graves to reinforce Bosnia's status as a "trouble spot" while positioning its orbital gaze as part of the mechanism of containment.

Because of its remoteness and abstraction, the satellite image functions as an overview of the war, and it draws on the discursive authority of meteorology, photography, cartography, and state intelligence to produce its reality and truth effects. Since it is digital, however, the satellite image is only an *approximation* of an event, not a mechanical reproduction of it or live immersion in it.[44] While the medium of television trades on its capacity for "liveness," which Jane Feuer and Patricia Mellencamp argue involves as much reenactment of the past and forecasting of the future as it does instantaneous access to unfolding events, the satellite image has an altogether different tense.[45] Because it is digital, its ontological status differs from that of the electronic image. The satellite image is encoded with time coordinates that index the moment of its acquisition, but since most satellite image data is simply archived in huge supercomputers, its *tense is one of latency*. Satellites are constantly and quietly scanning the earth, but much of what they register is never seen or known. The satellite image is not really produced, then, until it is sorted, rendered, and put into circulation, much like the film image does not really come into being until it is processed, projected, and seen. Satellite image data only becomes a document of the "real" and an index of the "historical" if there is reason to suspect it has relevance to current affairs. Unless the satellite image is selected and displayed, it remains dormant, gathered as part of an enormous accumulation of image intelligence that can be stored and used later.[46]

Archives of satellite image data thus create the potential for *diachronic omniscience*—vision through time—because they enable views of the past (and future with computer modeling) to be generated in the present that have never been known to exist at all, much less seen. Our understanding of the temporality of the satellite image should be derived through the process of its selection, display, and circulation rather than formed at the instant of its acquisition. Satellite image data can be sorted according to certain parameters and used to render a given place at a given time. It can be combined with imaging software and overlaid to form composites or animated to reveal transformations or movements in a place over a specified time. Or it can circulate raw—that is, as an abstract visual field that must be anchored and infused with meaning in order to signify anything other than its own orbital position.

When it becomes part of television flow, however, the satellite image imports its latency and remoteness into the coverage, recircumscribing the ways citizen-viewers are positioned to see and know the wars of distant others. The coverage of the Srebrenica massacre demonstrates Sam Weber's suggestion that television operates as "a screen which allows distant vision to be watched. . . . It screens, in the sense of selecting or filtering, the vision that is watched. And finally it serves as the screen in the sense of standing between the viewer and the viewed."[47] When it was televised, the satellite image of mass graves was itself carefully filtered and selected, and it, in turn, allowed distant vision to be watched. Remote sensing, in other words, became a televisual practice the moment it was embedded in and combined with a set of preexisting technologized looking relations.

American network television's circulation of U.S. satellite images of mass graves near Srebrenica was significant because it brought the orbital gaze into bold relief. The "broadcast" of a military gaze that is usually kept "top secret" exposed the violence of delaying and denying knowledge of events that are monitored (and produced) within a system of global military omniscience. The coverage was a product of the military-information-entertainment complex, which is a hodgepodge of conflicting agendas. Commercial broadcasters claim they have the right to profit by televising world events. Citizen-viewers insist they have the right to know about American intervention in the activities of other nation-states. The U.S. military wants to maintain what it calls its "Information Dominance" by continuing top-secret intelligence activities. These agendas of state intelligence, public knowledge, and corporate profit are constantly (re)negotiated in the kinds of war coverage that appear on American television screens. What is troubling is the way the intersecting agendas often work to negate U.S. accountability for what lies in its field of vision. In the case of the Srebrenica massacre, for example, the media can claim they were unable to access the region, and the military can insist that such events were unsuitable for television. Whenever it is strategic and/or appropriate, however, state intelligence agencies can and do share aerial and satellite perspectives, which are the foundations of a recent U.S. military campaign called Information Dominance.

The goal of Information Dominance was to enable Operation Joint Endeavor (the U.S.-NATO "peacekeeping" mission in Bosnia) to monitor occurrences within the war theater using real-time satellite and aerial perspectives and instantly relay that intelligence to a centralized command center. As Admiral Bill Owens explained, Information Dominance involves "the ability to see the whole battlefield at day [or night], in all weather, in real time."[48] As another officer put it, "The ultimate goal is to provide information that is seconds old, instead of hours old—virtually from the satellite, to the terminal, to the battlefield commander's hands."[49] Passing over Bosnia at the time of the Srebrenica massacre were highly classified Crystal satel-

lites that utilize the same KHII electro-optical systems on board the Hubble
Telescope. Their infrared capabilities enable them to take pictures in low
light and complete darkness, and they can deliver digital imagery of objects
as small as a grapefruit directly to a ground site in near real time. Another
top-secret satellite known variably as *Indigo, Lacrosse*, and *Vega* uses the
same imaging technology as the *Magellan* spacecraft, which mapped the
terrain of Venus.

The most advanced airborne system is the E-8C Night Stalker—a con-
verted Boeing 707 packed with state-of-the-art Surveillance Target Attack
Radar (STARS) equipment that flies over "areas of interest" and is "designed
to maintain night-time Information Dominance over Bosnia."[50] Its radar
can be programmed to "revisit" designated areas every thirty seconds and
record movement.[51] This aircraft system was used after the U. S. satellite
images were publicly released to monitor movements around mass-grave
sites in the northwest areas of Serb-held Bosnia. Also hovering over Bosnian skies were Apache helicopters, which "let gunners 'acquire' targets
by looking at them through an electronic monocle,"[52] and Predators, automated surveillance aircraft that can linger for days over given terrain and
transmit video live via satellite to ground stations throughout Europe.[53]
Much of the intelligence gathered by satellites and aircraft ends up in a
communications center called "Battlestar," which has been described by
designers as "a cross between the New York Stock Exchange and the Star-
ship Enterprise."[54] Information Dominance heightened the military significance of, and reliance on, satellite technology. As Major McElligott
boasted, "We can't physically be everywhere, but with satellites, we can get
a pretty good idea of what's happening everywhere. You don't have to be
there to see things. . . . It's the key to winning modern warfare."[55] As one
sergeant quipped, "Satellites are as essential to the soldier as an M-16 rifle
or a canteen."[56]

Despite the arsenal of vision machines controlled by U.S. "peacekeepers," the alleged execution of thousands of Muslims (and countless Serbs
whose deaths are rarely ever mentioned, much less counted) went on apparently unnoticed and uninterrupted. Crystals, Night Stalkers, Apaches,
and Predators crisscrossed Bosnian skies relaying image data to Battlestar
commanders, but U.S. officials claim not to have recognized the atrocities that unfolded in Srebrenica until weeks after they occurred. Satellite

images of impending mass executions did exist—recall the "before" and "after" images shown by Madeline Albright. Despite the U.S. military's claims to "information dominance" it apparently took weeks for the National Reconnaissance Office (NRO) computers to sort and process images that suggested these atrocities; thus they went on uninterrupted. What is tragically ironic here is that U.S. military technologies of omniscience appeared to have been blind to practices that the United States and its NATO partners were decrying as "ethnic cleansing." Intelligence officers in the NRO claimed to have had too much satellite intelligence to process, sort, and see.

The emphasis of real-time visibility and total knowledge of the war theater touted by officials of Information Dominance makes this act of "blind omniscience" especially violent and doubly invisible. For in a cultural climate in which truth is produced by hyperreal media technologies, an unseen event is bracketed off, denied entry into the "real." As Michael Ignatieff suggests, "Truth is always a casualty in war, but in virtual war, the media creates the illusion that what we are seeing is true. In reality, nothing is what it seems. Atrocities are not necessarily atrocities. Victories are not necessarily victories. Damage is not necessarily 'collateral.' But these deceptions have become intrinsic to the art of war. Virtual war is won by being spun."[57] The delayed release of satellite images of mass graves in Srebrenica points up the selective practices that undergird Information Dominance. It also challenges the assumption that the mere presence of the surveillance apparatus is enough to deter the most horrific acts of war. The State Department's public release of satellite intelligence and commercial television's subsequent repackaging of it contained the "liveness" of the Srebrenica massacre. The eventual circulation of these satellite images was calculated and strategic: it positioned the United States as a kind of "sweeper" coming in to clean up after the massacre, and it reduced the conflict yet again to a simple tale of Serbian bloodlust.[58]

A critique of the "coverage" of the Srebrenica massacre thus involves recognizing U.S. complicity with (or, perhaps better said, its "oversight" or "supervision" of) the atrocities that occurred there. It involves exposing the ways in which the United States, as an information superpower, monitors the traumatic experiences of grounded others from a distance and makes them known only when it is strategic to do so. Satellite images of mass graves are symptomatic of the American media's and military's reluctance to get "too close" to the war in Bosnia. When the war erupted in

1992, American television producers barely knew where Bosnia was on the
world map, much less how to cover the complex historical conflicts in the
region.[59] American political leaders were similarly unable to clearly define
their role in the UN's peacekeeping mission, until 1999, of course, when
the United States gave up on peacekeeping and decisively bombed Serbian
targets.

Despite the satellite image's inability to tell all, it circulates as if it does—
positioned as the most "objective" view, as a kind of official pipeline to
something happening half a world away. Still, given the overflow of satellite
image data acquired over Bosnia, we might ask what else did U.S. satellites see? Why were these satellite images selected for public release and
not others? While the United States has gathered a veritable flood of satellite image data over the Balkans throughout the 1990s, these were the
only images released because they allegedly revealed evidence of the mass
execution of Muslims in and around the UN-protected safe haven of Srebrenica. The circulation of the satellite images was not innocent, however.
It was key to U.S. political posturing as the U.S.-NATO alliance sought to
expose and condemn Bosnian Serb military aggression in the region while
simultaneously rationalizing its own failure to "keep the peace" in a UN-
protected safe haven.

So long as the public remains uninformed or misinformed about
the state and commercial practices of remote sensing, the military-
information-entertainment complex can use the satellite image's remoteness and abstraction, as well as the excessive volume of image data gathered
(which creates a slower sorting process), to deny knowledge of events that
are in full view. At the same time, however, media and military institutions can use satellites' constant scanning of the earth to produce what they
claim to be uniquely "objective" views of events whenever it is strategic to
do so. The satellite's gaze is trained on particular places for particular reasons, and, at the same time, it randomly acquires information about all
kinds of places for no apparent reason at all. Because of this, it can either
be mobilized as representing the ultimate authority of the state (and of
our unspoken faith in cartography) or as a completely abstract and uncertain point of view. It is the ambiguity between these positions that enables
the state to use satellite images in what Paul Virilio calls "strategies of deception."[60] Rather than provide evidence of massacre, the satellite images

of mass graves at once serve the seemingly antithetical strategies of denying knowledge and claiming omniscience. But if the strategies of deception are shifting, then so, too, must the practices of witnessing. Here the
citizen-viewer is invited to temporarily occupy the orbital gaze but lacks
the intelligence necessary to decode and interpret the view. We need forms
of witnessing that work within the context of virtual war.

Witnessing

Satellite images do not produce a total and omniscient register of earthly
activity. Rather, once publicly released, their emptiness and abstraction call
for—more than other forms of visual media—local validation and situated
knowledges. Albright's release of satellite images of Srebrenica triggered
an intricate sky-to-ground investigation to corroborate the orbital view. The
range of eyewitness accounts applied to the satellite image are far from
equal, however. Militaristic discourses privilege the dispassionate gaze of
the intelligence analyst and the anthropologist over the embodied experiences of victims and bystanders. The immediate accounts of eyewitnesses
on the ground were only treated as true after lengthy inquiry and verification. This hierarchy of discourses serves the military strategies of disavowal
that mark the atrocious event as past and, therefore, beyond intervention.
In this section I explore the possibility of using satellite images in ways that
challenge the state's hold on their range of possible meanings.

The broadcast of satellite images of Bosnia can be seen as part of a
broader shift in the coverage of war that began in 1991 during the U.S. war
against Iraq. As Douglas Kellner reveals, collaboration between the U.S.
military and media intensified during the Persian Gulf War as a response to
the antiwar sentiment fomented by TV coverage of bloody battlefields during the Vietnam War. To thwart viewers' opposition to U.S. attacks on Iraq,
TV cameras were pointed at Norman Schwarzkopf's "play-by-play" coverage instead of at Iraqi casualties and devastation. Commercial television
networks created euphemistic catch phrases for the war, such as "Show-
down in the Gulf" and showed lots of surgical strikes and little collateral
damage.[61]

The coverage of the Persian Gulf War actualized a set of synoptic relations between the military machine and the television camera, making
their gaze one in the same.[62] Kevin Robins suggests that the television

broadcast of missile-nose views of Iraqi targets "simulated a super-real closeness that humans could never attain. . . . The visible was separated from the sound and feeling of pain, from the smell and taste of burning and death. We could zoom in on the action, but the enemy remained a faceless alien. For us, the voyeurs, the reality of their deaths became de-realised."[63] If the coverage of the Gulf War detached the viewer from the reality (and materiality) of war, then satellite views of mass graves in Bosnia have further elaborated this process.[64] While the missile-nose perspective constructs an impossible proximity to the war event, the orbital gaze represents war at an impossible distance. In the satellite image of mass graves, the "faceless alien" is even further derealized, reduced to bits in a high-tech map, buried in the archive and then abstracted in the latent time and space of satellite image data. The satellite image, which according to proponents of Information Dominance is useful because it allows near-real-time visibility, appears on television as a memorial rather than a field of potentiality: The viewer is encouraged to see the Srebrenica massacre as having already transpired. Thus, as imaging technologies create the potential to move the citizen-viewer closer to a real-time experience of war, the scale of the image becomes more remote and its capacity for liveness suppressed.

We can no longer afford to distinguish such coverage of war from the art of war itself. As Ignatieff reminds us, "Wars fought on and for television—virtual wars—will not be less violent and destructive than those fought before the age of the television camera."[65] Further, we need to be able to imagine violence not only as mass execution but also as the practice of denying such an act, particularly when it appears in full view. Part of the violence of virtual war is in the U.S. military media's schizophrenic strategy of immersing the citizen-viewer within the exploding target, on the one hand, and detaching him or her from it in a seemingly serene orbital view on the other. Such alternating views structure a dialectic of anxious disorientation and pleasurable remote control for the citizen-viewer, generating a dangerous oblivion for those in whose names bombs are dropped and satellites launched.

In *Virtual War* Ignatieff asks the important question: "If war becomes unreal to the citizens of modern democracies, will they care enough to restrain and control the violence exercised in their name?"[66] If the coverage of American military interventions continues to place us in the cargo bays of fully computerized satellites and jets—incarnated as what Paul Virilio calls military "vision machines"—then we as citizens and viewers need to imagine ways to use them to bear witness to atrocities instead of embracing their seemingly "objective" and "omniscient" views. Put another way, if the American military-information-entertainment complex so forcefully regulates coverage of U.S. military interventions, then how can we appropriate and demilitarize the machines and screens that are mobilized to see for us? To answer this question I explore what it might mean to "witness" atrocities from the remote perspective of an orbiting satellite.

The satellite witness draws on the aesthetics of emptiness and uncertainty that are embedded within U.S. military representations in order to dislocate and destabilize their claims to "objectivity," "omniscience," and "truth." It attempts to demilitarize military perspectives and to see in them things that are not necessarily meant to be seen or known. In some cases this is the vantage point of the military gaze itself. Since satellite images are so abstract and remote, they necessitate closer scrutiny, re-viewing, careful description, and interpretation in ways that other images of war do not and often seem to "speak for themselves." By developing critical media literacies around the production and circulation of satellite images, we can begin a struggle to demilitarize orbital military platforms and use them to witness—that is, expose and deconstruct—strategic, technologized, and state-controlled media coverage of world events. The remoteness and abstraction of the satellite image, I argue, lends itself to a politics of witnessing, precisely because it must be grounded, situated, and infused with discourse if it is to signify anything other than the satellite's point of view. How might we use the satellite to bear witness not only to what happened in Srebrenica but more generally to hold the U.S. state and military accountable for what it sees and knows and how it acts (or does not act) based on that vision and knowledge? It's important to devise critical strategies for looking at the satellite image so that we don't simply adopt what Virilio calls an "eyeless vision"—an automated gaze that makes it all too easy to be complicit with the atrocities in view.[67]

Media studies scholars have theorized the practice of spectatorship, but few have considered how and when the spectator is positioned as a witness. Perhaps one explanation for this is that the definition of witnessing has itself remained somewhat fixed, essentialized as the physical bystander who happens to see an event unfold with his or her own eyes. In an age of technologized vision, however, we need to be able to imagine technologized forms of witnessing. In *Beclouded Visions* Kyo Mclear sug-

gests that "the scale and scope of modern warfare presents unique challenges to witnessing" and considers how it is possible, for instance, to witness an atrocity such as nuclear annihilation.[68] "Hiroshima and Nagasaki,"
she writes, "have produced a unique crisis of witnessing. . . . The atomic
bombings force an encounter with the limits of vision as a paradigm of
knowledge and ethics."[69] Indeed, how does the moment of nuclear explosion get visualized, much less witnessed? The dominant mode of representing the moment of nuclear explosion became the endless circulation of
a highly aestheticized mushroom cloud (as opposed to burned bodies and
scorched earth). To complicate the sky-based iconography of nuclear war,
Mclear turns to artistic practices, referred to as "countervisions," which
"lend human context to depictions that have tended to disembody, and,
hence, dehistoricize the effects of the bombs."[70] The art of witnessing, according to Mclear, is in the creation of art forms that generate alternative
discourses on traumatic pasts that are ultimately unrepresentable, as opposed to mass media coverage, which "tends to deny the historical conti-
nuity of events" and are "all too brief and promiscuous."[71]

While Mclear's definition of witnessing as artistic countervision is crucial, it is also limiting in its insistence that the witness exists only outside
of or apart from mass media culture. The effect of this distinction between
art and television (not just by Mclear but by cultural critics in general) has
been to reify the lack of radical potential within television (as both an institution and way of seeing and knowing) and to forgo struggles for progressive political expression on the airwaves. Mclear recognizes the need
to find ways of witnessing recent events in Bosnia and Rwanda but pro-
poses a kind of witnessing that exists outside of and apart from televisual
discourse. She writes, "Bosnian mass graves and Rwandan refugee camps,
crowded with the dead and dying, have begged for forms of witnessing that
might exceed televisually conceived understanding."[72] I want to define a
practice of witnessing that involves critical engagement with and accountability for the images of war that circulate *within* television culture—one
that is more about a struggle over television representation (and the way we
see and know) than about a struggle for an authentic viewpoint. Whether
we like it or not, the ways we come to see, know, and imagine the atrocities
of war are structured by media coverage, which is increasingly produced
as distant, abstract, and uncertain.

The satellite image produces an interesting paradox for the witness, for it brings all kinds of things into view, but its perspective is not just nonhuman but unearthly. What might it mean, then, to witness via satellite? Might it be possible to appropriate the remoteness and abstraction of the satellite image to, in effect, turn it upside down and invert its logic of omniscient detachment? This is what journalist David Rohde did on August 16, 1995 (six days after Albright's announcement), when he used a "blurry, faxed copy of a US spy satellite photo" to search for the alleged mass graves near Srebrenica. Rohde wondered if there were really mass graves as U.S. officials alleged from the photo.[73] When he arrived at an empty field where the ground had recently been turned, he saw a decomposing human leg protruding from the dirt and large empty ammunition boxes scattered across a field where cows and horses now grazed.[74]

Rohde also discovered the tattered remnants of personal identification once carried by the Muslim men whose bodies lay in the earth beneath him. High school diplomas, photographs, and handwritten notes littered the fields. Rohde witnessed the minutiae that the satellite could not pick up—washed-out personal photographs of a Muslim youth from the village of Kravice, handwritten notes from a March 1995 meeting in the village of Potocari with instructions about how to load weapons, a 1982 elementary school diploma, Muslim prayer beads, fabric and clothing, receipts, and election ballots from Srebrenica.[75] Near the soccer field in Nova Kasaba, Rohde entered the abandoned building allegedly used to imprison and torture Muslim men. He discovered human feces, blood, and bullet holes all over the walls—evidence that large numbers of persons had been confined and perhaps shot there. Rohde also noticed truck and bulldozer tracks leading to suspected mass graves.[76]

Since the satellite image privileges geography over what Jennifer Gonzalez has called "autotopography"—the small personal elements that "form a visible and tactile map of subjectivity"—its status as an overview represents the potential to erase the minute traces of human life and, quite literally, bury the subject.[77] Rohde's accounts infused the satellite image with personal belongings or "forms of self-representation," which Gonzalez claims "anchor the self-reflective image of the subject within a local, earthly cosmos."[78] Rohde excavated the "autotopographies" of Muslim men as he wandered the fields around Srebrenica, and his report initiated a

A girl's lost shoe settles
among the personal ruins
and autotopographies that
orbital vision cannot de-
tect. Photo by the author.

struggle to make the satellite image witness—that is, to see and read the
satellite image differently than did military officials.

By seeing the satellite image as a site of activity rather than a memorial,
Rohde recodes satellite intelligence as a field of public inquiry, inscribing
the subjectivities of Muslim victims into the image and inviting further
discussion. Roland Barthes offers the term *anchoring* to refer to the way
that meaning is attached to an image (such as a caption to a news photo) to
make it signify in a certain way. Anchoring has the effect of restricting the
range of possible meanings by encouraging the viewer to adopt what is pre-
scribed.[79] In a sense I am arguing for a reversal of this anchoring process,
where the witness, instead of accepting the state's attempt to anchor the
meaning of the satellite image, seizes its emptiness and abstraction as im-
petus to infuse it with partiality, situated knowledges, and local tales. The
satellite witness engages, in other words, in a practice of *semiotic infusion*:
The orbital gaze is turned on its head (so to speak) so its claims of military
intelligence can be fleshed out. Simply put, it involves a more literal interpretation of the term *remote sensing*, exploring how the senses, the sensed,
and ways of making sense are related to orbital vision. The satellite witness
subjects the military intelligence apparatus to the same kind of scrutiny
and interpretive analysis that it for decades has applied to the world.

Significantly, Rohde's journalistic account of the Srebrenica massacre
was subjected to skepticism until it could be confirmed by more "disinter-
ested" or "neutral" U.S., NATO, and UN investigators and forensic anthropologists. What I want to highlight, however, is Rohde's refusal to accept the
satellite image as evident. He recognized within it the distinction between

the orbital gaze and the potential of his eyes to see differently. This recognition coincides with Kaja Silverman's insistence on the disjuncture between the apparatus/gaze (as a set of historically situated material practices) and the eye (as a sensory organ). Silverman encourages us to conceptualize the apparatus/gaze as distinct from the look: "The look has all along possessed the capacity to see otherwise from and even in contradiction to the gaze. The eye is always to some degree resistant to the discourses which seek to master and regulate it, and can even on occasion, dramatically expose the representational logic and material practices which specify exemplary vision at a given moment in time."[80] The look does not simply take in the satellite's view but acts on it and in relation to it with the sensory processes of the eye.

The satellite witness recognizes the ambiguous allure of the orbital view and can use it to expose and attract attention to abuses of power and calculated denials that emanate from a position of seeing and knowing too much.[81] Satellite witnessing involves exposing information-based, simulated, and composited forms of violence that are often trying to hide themselves or pass in seemingly innocent discourses of "monitoring," "peacekeeping," or "protection." This witness presumes there may be something to see in even the most obscure view and as such is positioned similarly to Sam Weber's television watcher: "To watch carries with it the connotation of a scrutiny that suggests more and less than mere seeing or looking at. To watch is very close to watching out for or looking out for, that is, being sensorially alert for something that *may* happen. . . . It involves watching out for something that is precisely not perceptible or graspable as an image or a representation. To 'watch' is to look for something that is not immediately apparent."[82] The more remote and abstract the satellite image, the more inquisitive the eye, ever aware of the latent temporality of its production and the selectivity of its circulation.

The broadcast of satellite images of mass graves in Srebrenica provided an opportunity for citizen-viewers to differentiate their look from the gaze of U.S. military intelligence. Watching these satellite views on TV involves the potential to commandeer them, to energize and activate what is missing in them, whether the bodies of Muslim men or the U.S. military's own "oversight" of the massacre.[83] Unless citizen-viewers object, satellites will continue to see/know for us, interpellating us as distant monitors and remote controllers of others' atrocities and traumas. What is ultimately important about the oppositional potential of the look, Silverman suggests, is "its capacity to intervene within the field of vision. . . . It underscores the capacity of the look not merely to see what is inapprehensible to the camera/gaze, but to alter what that apparatus 'photographs.' "[84] Perhaps the look of the witness can help to demilitarize the satellite image.

Though the look has an oppositional potential, there is no guarantee that what it sees will necessarily be more clear or pure than what the satellite sees. The satellite witness, in other words, is not searching for total truth or crystal clarity but rather embraces a fuzzy logic of representation, insisting on the irreducibility of war. This position of witnessing might be read as patently antithetical to the obsessively realist special effects seen in Hollywood war films such as *Saving Private Ryan* or *Pearl Harbor*. Perhaps ironically, the military satellite image of mass graves functions as an important counterpoint to these glossy, nostalgic, and nationalist films. Its abstraction reminds us that there is no way that the image itself—whether in the form of digital satellite intelligence or filmed reenactment—could ever fully encapsulate the lived microphysics or the psychic life of war. At the same time, however, it is *in the image* that citizen-viewers must struggle for the autonomy of the look to maintain its separation from the orbital gaze of the military intelligence and the military gaze of filmed entertainment.

Satellite witnessing is a critical practice that refuses to accept the satellite image as an omniscient view, a strategic map, a final perspective and instead appropriates its abstraction to generate further interrogation, discussion, and inquiry. Satellite witnessing is audacious in that it imagines military image data not as dormant state property but rather as a volatile discursive field that can be used by citizen-viewers to expose, question, and critique military techniques of observation, intervention, action, assistance, or "peacekeeping." We need to infuse military satellite images with public debates, countervisions, and situated knowledges precisely because they appear so far removed from spheres of civic and public life and so far removed from public responsibility and accountability.

Satellite images are fascinating as sites of meaning because they can at once reveal and conceal events that we cannot bear to look at. Their representational codes seem to negate the perspective of the witness altogether. Part of the problem, however, is that we have historically expected

the witness to provide comprehensibility and verification. We need to recognize that in the information age, the function of the witness changes:
He or she might be exposing misinformation, pointing the finger at a *partially* responsible party, exposing the violence and oppression of the gaze/
apparatus itself, or simply holding onto an unanchored point of view. In
addition to implying responsibility for what one sees, witnessing involves
recognizing horrific acts that are meant not to be seen. Witnessing in the
information age involves being able to notice atrocities that might not even
be imagined as part of the realm of the visible, things on the "threshold of
the visible."[85]

There can be as much violence in the use (or misuse) of images and information mobilized to regulate public knowledge of events as in acts of
physical violence themselves. The satellite views of mass graves in Bosnia
amount not only to an abstraction of the Srebrenica massacre but, more
pointedly, to a kind of *discursive antiseptics* in the way the U.S. military
media televised it from orbit. These antiseptic views position the U.S. intelligence apparatus not as a benign observer but as a culpable partner in
the very practice of "ethnic cleansing." In 1996 and 1997 television networks aired segments featuring forensic anthropologists excavating the
mass-grave sites, suggesting the West had come in to "clean up" the massacre, further compounding the meanings of *ethnic cleansing*. This layering
of antiseptic discourses suggests the term *ethnic cleansing* should refer not
only to the military purging of ethnically defined civilian populations based
on nationalistic impulses but also to the erasure, suppression, or refusal of
the politics of ethnic differentiation within the televisual coverage of the
war. In these satellite images not only are the bodies of Muslims conspicu-
ously missing, but so, too, are those of Serbs and Croats.

These orbital views serve as a cogent symptom of U.S. culpability precisely because they expose how the state's gaze averts, reduces, and con-
tains practices of "ethnic cleansing" while at the same time proclaiming
to "discover" them. Perhaps what is also being indexed in these images
is the U.S. unwillingness to come to terms with its own ethnic and racial
politics as it monitors and confronts those of others. Boundaries may be
more fluid in the postcommunist era, but satellites continue to draw lines
between East and West. There remains a kind of iron curtain visuality in
place, which is perhaps most perceptible in the (over)production of the
military satellite image, itself a legacy of cold war geopolitics. The mass

A fallow field across from the Potocari battery factory becomes a site of unframed earth. Photo by the author.

circulation of these satellite images thus serves a broader strategy of Western disavowal, detachment, and denial by symbolically burying not just the Muslim victims but the bodies and histories of all those involved in the Bosnian conflict (as well as Yugoslavia's communist past), collectively laying them to rest in a digitized mass grave.

Conclusion

In April 2001 I went to Bosnia to shift my position, to move my eyes from orbit to the ground. While I was there, Slobodan Milosevic was arrested at his compound in Belgrade, Croat nationalists were threatening to withdraw from the republic of Bosnia-Herzegovina, Macedonian troops were bombarding Albanian rebels near Skopje, and the Muslim women of Srebrenica were trying to erect a memorial to their husbands, sons, and brothers killed in July 1995.[86] In Srebrenica a UN translator escorted me to places referred to by UN workers as "all the tragic sites," which suggested that an unlikely form of "genocide tourism" has taken shape there. He took me to the Potocari battery factory, where Muslim women and children were separated from men who were allegedly trucked away and killed, and the fallow field across the street that is the proposed Muslim memorial site. I walked through Muslim warrior Naser Oric's abandoned mansion overlooking the village (where the region's Muslim militia organized and women were reportedly held as sex slaves) and drove through plum tree–covered hillsides that were commanded by Dutch UN troops before being overtaken by the BSA. In Café Kum I encountered a former Serbian military officer who, I was told, had recently been indicted by the War Crimes Tribunal. The mass-grave sites themselves were off limits. And when I asked about the possibility of seeing them, I encountered responses that ranged from dead silence to a suggestion that I read a book about the massacre. Another person explained that the grave sites were simply too difficult to get to.

There is a code of silence in Srebrenica that is difficult to permeate, especially for an outsider like me. Even on the ground there are forms of screening, filtering, and sorting, forms of television. This experience of *distant vision up close* emerges not just as an issue of translation or tightly held secrets; it is related more to a sense of betrayal in the way the West used Srebrenica during its most vulnerable moments, the way the West buried Bosnians in full view. I noticed this most acutely while I was shooting video footage in Srebrenica. Even though I may have imagined my look as distinct from the gaze of U.S. military intelligence, I realized that Bosnians do not necessarily see it that way. The look of an outsider feels penetrating and cruel to those who want to call this place home.

What was once a UN "safe haven" is now what UN workers call a "DP camp." Many of the people who live in Srebrenica are Serbs from the krajina (rural regions) who were displaced during the war. Most of the village's Muslims moved away to Tuzla or Sarajevo after 1995, but many plan to return home.[87] Even though most of Srebrenica's residents did not live there in July 1995, they, too, remember the violence of being uprooted from their homes and seeing family and friends killed. Their traumas, however, were not necessarily monitored and made known by satellite vision. Those who have settled in Srebrenica are reminded that it is a "landscape of atrocity," not only by the perpetual flow of UN vehicles, SFOR (Stabilization Force) troops, and civilian forces assigned to manage and police the region, but also by journalists who tell the world this "peaceful valley is actually a graveyard."[88]

Rather than imagine Srebrenica as a graveyard, I prefer to think of it as a site of *unframed earth*. We forget in the age of virtual war, digital archives, and imaged memory that the materiality of war is still stored in the soil;

it coagulates in the crust of the earth and sediments in the skin. This site
was as abstract to me up close as when I first saw it on television, scanned
and layered in a series of mediated frames. Witnessing became a *fantasy of
proximity.* The main difference is that while standing in Srebrenica, I recognized the place as unframed earth, a site unscreened. Each time a satellite
passes overhead, someone or something is framed. In her essay "Starting
from Scratch: Concepts of Order in No Man's Land," Cornelia Vismann
reminds us that order emerges as a spatial concept: "Order, the universal
measurement of land and the specific order of the soil, starts by drawing
a line." "Order becomes concrete and dynamic," she writes, "immediately
responsive to war's marks in the soil."[89] In Bosnia soldiers drew lines in
the soil with the movements of their bodies, tanks, and gunfire, and victims drew lines with their flesh and blood. But U.S. satellites drew lines
there as well, lines that cut just as deep in the soil as the bulldozers that dug
mass graves. Satellites framed the places and people of Srebrenica, regis-
tering them as bits of time and space, demarcating the global viewer from
the grounded viewed. Unframing this piece of earth involves exposing the
satellite's invisible carving of the planet and being able to imagine the lines
that it draws as open wounds.

The satellite witness shares more than a passing resemblance to the
genealogist since both are engaged in processes of getting closer, examining surfaces of the particular, while remaining alienated and removed.
This structure of feeling, this fantasy of proximity, also resonates with descriptions of television as a medium of intimacy, as one that transmits
from a distance material that is framed close-up. If television can be understood in part as a dialectics of distance and proximity, then it would seem
that its convergence with the satellite might further extend rather than
foreclose possibilities for seeing and knowing *difference from a distance.*
Perhaps we could imagine the satellite as generating a kind of "orbital
pull," a metaphorical dislocation, a figurative removal from the zones of
security and comfort in the world, forcing us to recognize the partiality of
vision and knowledge and to embrace the unknown. We need to devise ways
of seeing and knowing difference across distances that complicate rather
than reinforce militaristic and scientific rational paradigms. At the same
time, however, we cannot be naive to such tendencies as they suffuse the
televisual.

Egypt and Qait Bay from the perspective of an astronaut
in orbit. Courtesy of Earth Sciences and Image Analysis.
Laboratory, NASA Johnson Space Center.

Satellite Archaeology

REMOTE SENSING CLEOPATRA IN EGYPT

In 1963, just after the first experimental TV relays via *Telstar*, an Egyptian diver swimming in Qait Bay discovered the ruins of ancient Alexandria, once inhabited and ruled by Queen Cleopatra. Not until three decades later in 1993, however, did French archaeologists Jean Yves Empereur and Frank Goddio seek funding from the Hilti Foundation, France, and Egypt to excavate the underwater site. Since then, Goddio and a team of archaeologists have used remote-sensing and global-positioning satellites to support their excavation of Cleopatra's palace. In 1996 Goddio's team began to present its ancient findings to the world in television news reports, documentaries, and Web sites. This chapter delineates how satellite television has been used to locate, sense, and map Cleopatra's underwater ruins.

Whereas chapter 3 examined how military and media institutions used a specific configuration of satellite television to regulate knowledge of wartime atrocities from afar, this chapter considers how the technology is imagined as extending the gaze of archaeologists in search of ancient civilizations. This shift from mediated war to mediated archaeology is symptomatic of an attempt to blend social constructivist and genealogical approaches to study various arrangements of "satellite television" technology. Combining these approaches involves not only exploring how these technologies may take shape and be put to use in different ways but also setting those different uses into relation with one another. Such a critical practice emphasizes the multivalent potentials of technologies whose physical properties may be similar but whose uses and effects may differ dramatically either because they emanate from a different set of institutional conditions or because they engage different regimes of knowledge, power, and truth.

Together, chapters 3 and 4 explore how "satellite television," defined here as remote-sensing practices, has been used to extend and maintain epistemological boundaries of Western Eurocentrism. Just as Srebrenica has been constructed as a site of civilizational tension between the communist and Islamic East and the democratic West, so too is the figure of Cleopatra imbricated in similar conflicts, perched on the fulcrum of Western (Greco-Roman) and Eastern (Oriental/African) legacies and competing origin myths. Whereas in the case of the Srebrenica massacre satellite images were used to represent wartime atrocities as occurring in the past, in Alexandria satellite use has become part of a broader effort to revivify and affirm historical narratives of Western civilization.

If, as Michel Foucault suggests, archaeology is more about discourse and power in the present than it is about the truth of the ancient past, then there is much to explore in the excavation of Cleopatra's palace. Using remote-sensing and global-positioning satellites, archaeologists have intensified an already existing scientific and cultural fascination with the ancient Egyptian queen, pinpointing her private living quarters, lost underwater for centuries. To study the formation of a discursive object such as Cleopatra is, according to Foucault, to "map the surfaces of . . . [its] emergence."[1] This chapter explores Cleopatra's emergence in the surface of the satellite image and the GPS map to energize a constellation of cultural imaginaries and scientific pursuits.

In this chapter I first consider archaeology as a televisual practice. Such a notion may seem unlikely since archaeology involves digging in the dirt and television involves sending signals through the air, but both are practices of remote sensing. Instead of seeing across vast distances, the archaeologist peers back in time, reading the earth's surface as a text. Where television represents remote parts of the world in all its vitality, satellite images frame the earth as a massive excavation site waiting to be plumbed. This comparison is intended to provoke further discussion of the televisual's permeation of scientific disciplines, particularly those organized around practices of distant observation and discovery. To develop this point, I discuss the use of remote-sensing and global-positioning satellites in the excavation of Cleopatra's palace in Alexandria. Rather than embrace the technologically determinist claim that satellites simply enhance and improve archaeologists' vision, I suggest that the "scientific" use of satellites must be understood in relation to "cultural" discourses that construct Cleopatra

as a sexual spectacle, site of racial ambiguity, and monument of Western civilization. As Barbara Holland suggests, "Cleopatra's looks are one of the burning issues of the ages."[2] Historians, writers, artists, and filmmakers have imagined her as a virtuous suicide, inefficient housewife, exuberant lover, professional courtesan, scheming manipulator, femme fatale, and rambling bimbo.[3] The intensity and persistence of scientific and fictional fascination with Cleopatra exists in proportion to the sexual and racial enigmas that she evokes and represents. The final section of the chapter thus examines discourses surrounding the excavation of Cleopatra's palace, which relocate her mysterious and excessive sexuality underwater, embodied in statues, busts, and treasures that are eventually thrust through its glassy surface. The remote sensing of Cleopatra, then, is not just a technologized archaeological excavation; it involves the compulsion to imagine the ancient queen as a sexual spectacle that can be sensed, materialized, and touched in the present using satellite and computer vision. Once again, the structures of feeling elicited by remote sensing are contradictory, but here the *temporal* distance of the gaze intensifies the desire for proximity, closeness, and even tactility. As we saw in chapter 3, the more abstract, remote, and uncertain the satellite image, the more tightly it is tethered to discourses of the body and the senses. Like the satellite witness, the archaeologist has a fantasy of proximity, but this one is as much about the Western imaginary's inability to reconcile its ambivalence toward feminine sexuality, power, and authority as it is about the "discovery" of ancient ruins. That is, the use of global-positioning and remote-sensing satellites in this particular excavation can be seen as symptomatic of a voyeuristic attempt to uncover and fix Cleopatra's identity once and for all and to affirm her status as a figure of the West.[4] Despite the objectivist claims of science, the cultural imaginary will also inform what the world searches for and finds in her underwater palace.

Archaeology and the Televisual

Archaeologists' use of satellites to spot ancient artifacts is somewhat ironic, given their notoriously tactile connection with the physical ruins of bygone eras—their commitment to the notion that the truth of ancient times can only really be found by digging in the dirt. But during the past two decades archaeologists have relied increasingly on satellites and computers to locate traces of ancient civilizations. Just as the satellite has become an important technology of modern warfare, so, too, has it become "a common part of archaeology's tool kit."[5] As one writer puts it, "The picks and shovels of traditional 'dirt archaeology' are giving way to proton magnetometers, ground-penetrating radar and infrared sensors."[6] Archaeologists began using remote-sensing satellites to survey landscapes during the mid-1980s, after having used aerial photographs taken from balloons and aircraft for decades.[7] But aerial photographs could not "see" beneath the earth's surface and thus could not make buried ancient matter visible to the human eye. *Landsat* images released in the 1980s not only presented broad regional perspectives, but their infrared and false color composites revealed subsurface layers of the earth. As NOVA puts it, "Landsat revealed a psychedelic world of magenta vegetation, black water and gold forests."[8]

Tom Sever was the first archaeologist to establish a working relationship with NASA during the early 1980s. His use of satellite images to reveal ancient roadways in Chaco Canyon, New Mexico, convinced NASA officials that space technologies could have a dramatic impact on archaeology. One NASA writer insisted, "Sever's results changed archaeology forever. . . . Buried prehistoric walls, buildings, agricultural fields and roadways stood out as if daubed in paint."[9] Following Sever's lead, archaeologists began to rely more heavily on remote sensing, especially with the installation of multispectral and infrared scanning devices that allowed them to see a wider range of the electromagnetic spectrum.[10] In 1992 the *New York Times* featured a full-page illustration of the remote-sensing process, declaring "Distant Eyes Bring Prehistory Closer."[11] Satellite sensors, the article explained, "see" far beyond the narrow range of the visible light spectrum (the wavelengths between ultraviolet and infrared frequencies) to which the human eye is limited.[12] Whereas some optical sensors identify surface objects by analyzing reflected light, microwave, infrared, and radar-imaging sensors can pierce clouds, jungle canopies, sand, and even soils. As *New Scientist* claimed, "The human eye is no longer good enough for the discriminating archaeologist. Where you or I might see an ordinary field of tall grass and trees, an archaeologist knows that beneath it may lie a Paleolithic village, undisturbed for thirty centuries."[13] Another writer explained, "A satellite's bird's-eye perch gives archaeologists an opportunity to discover very large patterns and features in the landscape that are all but imperceptible from the ground."[14] This ability to " 'see' the glow of heat energy given off by the ground" reportedly augured "a revolution in archaeology that has brought a new thrill of discovery to the old trowel-and-shovel discipline."[15] No longer only digging in the dirt, archaeologists have used the satellite's colorful abstractions of light, heat, and matter as tools of excavation.

Sometimes anomalies in the satellite image lead directly to a site of antiquity, for ancient sediments may have a characteristic infrared signature that sets them apart from more modern terrain.[16] Minute differences in heat retention can point to filled-in irrigation ditches and entrances to tombs or temples, which are often backfilled with loose dirt. The Thermal Infrared-Multispectral Scanner (TIMS) scans energy waves emitted in the infrared and radio bands, and it has been particularly useful for finding remains long buried just beneath the surface of the earth.[17] The TIMS could, for instance, distinguish the wavelength or "signature" of a specific tree species such as Ramon, which is often associated with Mayan ruins. Another device commonly used by archaeologists is the Calibrated Airborne Multispectral Scanner (CAMS), a nine-channel instrument that scans the visible and near infrared. In 1997 NASA launched a new multibillion-dollar series of satellites and instruments, known as the Earth Observing System (EOS). The EOS doubled the daily production of raw satellite data and will generate global environmental remote-sensing data for fifteen years.[18] The goal of this system is reportedly "to read even more closely the script of the Earth's surface."[19]

Such references to the surface of the earth as a "script" highlight the interpretive nature of remote sensing. Here, too, as in the case of Bosnia, the satellite image must be grounded—that is, read, decoded, and contextualized—in order to signify anything other than its orbital perspective, to even remotely make sense. The satellite image frames the earth as a semiotic field that is constituted not only by processes of geologic sedimentation and atmospheric conditions but by accumulations of cultural meaning as well. On the one hand, the "earth as script" metaphor exposes the fact that satellite-based archaeology is enmeshed with cultural practices of interpretation. On the other hand, the metaphor is problematic because it implies world history can be written or represented without an author and that the earth itself has a naturally evolving narrative whose self-evident truths need only be uncovered or unearthed. Tom Sever represents such a position:

> As a species, we've been literally blind to the universe around us. If the known electromagnetic spectrum—from cosmic rays to visible light to huge seismic waves of the earth's interior—were scaled up to stretch around the planet's circumference, then the human eye and conventional film would see only the visible light portion, equal to the diameter of a pencil! Our ability to build detectors that see where we can't see and computers that bring invisible information back to our eyesight will contribute to our survival on Earth and in space.[20]

The invisible information gathered by remote sensing is treated as an antidote to human myopia. As if it can penetrate the artifice of human culture itself, the remote-sensing satellite is invested with the ability to read "the script of the Earth's surface" in its native tongue.

In treating the earth's surface as a script, archaeology imagines the planet as the raw material of the ancient past. The satellite image is used to plumb its layers for meaning and can be used to produce a network of historical narratives from dispersed and remote locations. As one writer announces, "Remote sensing devices are forcing the earth to give up its secrets."[21] Consider, for instance, the multitude of excavation sites seen with satellite technologies: buried footpaths in Costa Rica leading to villages and communal cemeteries; the quarries and citadels of Hellenic Greece; fossilized hominids in Ethiopia's Great Rift Valley; Nikopolis, a Greek city built to commemorate Octavian Caesar's victory over Antony and Cleopatra; the ancient city of Ubar, known as "the Atlantis of the Sands"; the Giza pyramids of Egypt; prehistoric Mayan cities in the Yucatan jungle; the British warship *Mary Rose*, from the era of Henry viii; *L'Orient*, flagship of the fleet that carried Napoleon to Egypt in 1798; Fort Griswold, a Revolutionary War battlefield in Connecticut; forty-three thousand seventeenth-century Dutch artifacts lying beneath New York's Wall Street; the tomb of Genghis Khan; the walls around the ancient city of Troy; a long-extinct river in the sands of the Sahara; and the Silk Road, an overland trading route used by Marco Polo and other travelers between Europe and the Far East.[22]

The practice of satellite archaeology shares a similarity with early satellite spectaculars like *Our World*. Just as producers of the first live satellite television shows scanned the earth to bring the world into view, so, too, have archaeologists scanned its surface to locate buried treasures. Instead of producing the world as a "global now," archaeologists use remote sensing to assert the immediacy of the ancient past. And instead of approaching the world as a panoply of unfolding events, the practice of satellite archaeology understands it as a repository of invisible artifacts simply awaiting discovery and historical narratives simply waiting to be written. As Sever insists, remote sensing "makes the invisible visible, the hidden found."[23] Here the satellite becomes a technology of scientific objectivity and positivist logic: The standby mode of live global television is rearticulated as a practice of revivification. Whereas *Our World*'s remote cameras produced the unfolding time and space of the global present, cameras and scanners on board remote-sensing satellites acquire data used to conjure the ancient past. As Jody Berland explains, "Satellite-based observational technologies . . . render the earth's surface, together with its surrounding atmosphere as usable visual information. . . . They create images or surfaces out of materials which were not previously conceived as visual."[24]

What I want to explore further here is the idea that archaeology can be understood as a televisual practice. Not only are archaeologists engaging in a practice of "remote seeing" as they look from orbit and peer through centuries; their gaze is directed to locate material for the production of historical narratives. In other words, satellite images of the earth give rise to various episodes in the history of (Western) civilization, episodes that become sites of excavation extracted for maximum return. These historical episodes are not so much found as they are formed. Matters of the past are always interpreted through lenses of the present. Like television producers, archaeologists must establish their practice in relation to the present. And like a television series, the artifacts they "discover" often become commodity fetishes that circulate in a system of cultural and economic exchange. The excavation establishes an episodic pattern because the site must be returned to again and again until its enigmas can be sufficiently resolved. Further, in the case of the excavation of Cleopatra's palace, news reporters and archaeologists even unconsciously allude to generic elements of the soap opera in their framing of the project, repeatedly char-

acterizing the Egyptian queen as a sultry seductress and the underwater
site as her ancient trysting ground.

I draw these parallels across the discourses of archaeology and the televisual not as whimsy but to complicate the positivist claim that satellites
have simply extended the archaeological gaze, making the practice more
precise, efficient, and objective. What I want to suggest, instead, is that
positing archaeology as an analogue to the televisual may be a way to challenge and restructure its guiding logic of scientific objectivity. For as John
Hartley reminds us, "television is irreducible to science."[25] The televisual
(in whatever form) is present in more ways than most care to acknowledge.
Recognizing it as such, particularly in disciplines whose epistemologies are
based on notions of distant observation and discovery (such as history, archaeology, geography, meteorology, and astronomy), may help debunk the
claim that the scientific gaze can be free of culture. Scientific knowledge is
necessarily constituted through and governed by a variety of cultural logics
and practices—and the televisual is certainly one of them.

Archaeologists' interest in efficient and "noninvasive" (or electronic)
practices suggests there may be further parallels between archaeology and
the televisual. Satellites are understood by archaeologists as making the
practice of seeing through time and space much more efficient. As Sever
states, "In one satellite sweep, we can survey a big river valley that would
take an archaeologist a lifetime to crisscross and excavate on the basis of
hunches."[26] Paleoanthropologist Tim White describes the satellite as "a
navigational tool" that enables archaeologists to "turn enormous areas into
manageable areas."[27] Finally, Anna Roosevelt, an archaeologist working in
the Brazilian Amazon, notes, "Most archaeologists just start digging. It
might take ten years to do a big site." She then asks, "Why not survey it geo-
physically in two weeks?"[28] By framing large areas of the earth, the satellite
image again becomes an encapsulation, but in this case it functions either
as an overview revealing the earth's hidden secrets or as a master shot an-
ticipating further action and closer scrutiny. Just as live satellite television
efficiently packages the world for viewers, so does remote sensing perform
such a task for archaeologists.

Some archaeologists claim that new satellite and computer imaging
technologies allow them to "pick the sites apart electronically" and avoid
damaging the ruins. Such "noninvasive" archaeology, as it has been called,

enables archaeologists to allegedly "map the past without disturbing it."[29]

In some cases electronic and digital excavation techniques are used to avoid
digging altogether.[30] Before excavations, some archaeologists create visual
composites of the sites using computer-aided design and Geographic Information Systems (GIS) software. Developed in the mid-1980s to translate data from maps or remote-sensing instruments, GIS provides interactive databases and images that can be manipulated on a computer screen.
In their efforts to reconstruct Nikopolis, for example, archaeologists recorded the coordinates of above-ground monuments, transferred the data
to a computer-aided design program and generated 3-D digital composites
of the entire site.[31] Scientists have also created 3-D reconstructions of el-
Qitar, an ancient mountaintop fortress overlooking the Euphrates River in
northern Syria. In Alexandria, Goddio has used GPS to map the longitudinal and latitudinal coordinates of artifacts and then combined this data
with satellite images and computer software to generate accurate 3-D interactive maps of the site. These high-tech maps, explains *Technology Review*,
are "more accurate than the rough, often incomplete sketches traditionally
made by pacing off distances."[32] Satellite images and computer simulations
are thought to generate a more "authentic" historical record than the mea-
surements and calculations of archaeologists working on the ground. The
notion that these images and simulations "map the past without disturbing
it" is a myth, however. Whether the past is probed by a satellite, an archaeologist, or a filmmaker, it is not simply "there" waiting to be found; rather,
it is actively assembled and made.

While the comparison of archaeology and the televisual may seem uncanny, it is intended to be. I want to consider the televisual not only as a set
of historically situated industrial, textual, and reception practices but also,
more broadly, as epistemological practices that structure worldviews and
knowledges from a distance. Such an approach allows us to consider how
the "logics of television" have been unknowingly embedded within various
disciplines and forms of knowledge and power in the late twentieth century
and the early twenty-first. If, as I suggested in chapter 1, television works to
extend and naturalize itself as a form of global presence, then certainly it
would not stop at the doors of disciplinary boundaries. The televisual is an
epistemologic system that intersects with and permeates various forms of
knowledge and power, especially those of scientific fields that pride them-

selves on masterful seeing and knowing the world from a distance.[33] The relationship between archaeology and the televisual is not just a critical fiction either. One need only turn on the Discovery Channel, the Travel Channel, the History Channel, PBS, or NASA-TV to find a more literal commingling of the two. The excavation of Cleopatra's palace has itself resulted in a lucrative media synergy between Frank Goddio, the Hilti Foundation, and the Discovery Channel, wherein archaeology quite literally becomes television entertainment.[34]

What matters here, however, is the ability to use the televisual as a discourse with which to critique and dislocate archaeology's claims to "better" scientific objectivity and historical accuracy thanks to satellite imagery. If, according to Foucault, archaeology involves questioning the very processes by which knowledges are formed in the first place, then certainly satellites are technologies of archaeology in this broader sense, for they open up discursive fields as much as they resolve them. In *The Archaeology of Knowledge* Foucault draws a useful distinction between "scientific domains" and "archaeological territories." While conventional archaeology might describe itself as a "scientific domain," Foucault's "archaeological territory" refers to the interplay between culture and science—its boundaries are not so clearly delineated, and it encompasses "literary" and "philosophical" texts as well as scientific ones:

> What archaeology tries to describe is not the specific structure of science, but the very different domain of knowledge. Moreover, although it is concerned with knowledge in its relation to epistemological figures and the sciences, it may also question knowledge in a different direction and describe it in a different set of relations. It is by questioning the sciences, their history, their strange unity, their dispersion, and their ruptures, that the domain of positivities was able to appear; it is in the interstice of scientific discourses that we were able to grasp the play of discursive formations.[35]

The reach of scientific inquiry goes far beyond the specific empirical object at hand. For while the "scientific domain" (or that space which is defined by the academic discipline of archaeology) of the Cleopatra site is bounded by the contours of the Alexandria harbor, the "archaeological territory" of the "dig" spans continents and centuries. It percolates through long-held assumptions about Western civilization and its origins, through

nineteenth-century scientific racism and the modern nation-state, and through a twentieth-century popular culture fascination with the ancient queen's sexuality. We might consider, then, what happens when satellite excavation and the scientific gaze are trained on an object that has been imagined within and popularized by cultural discourses. Certainly in the case of Cleopatra the "natural" script of the earth's past has had a coauthor: the invisible hand of objective truth has been guided by the more shaky hand of two millennia of cultural reinvention. How, we should ask, do cultural discourses shape the gaze of the scientist engaged in the practice of satellite excavation? With this in mind I would like to move the scientific domain of archaeology in a different direction and into yet another set of relations. In addition to considering archaeology as a televisual practice I want to explore how its scientific gaze is formed in relation to the history of cultural fascination with Cleopatra that occurred during the twentieth century and continues into the twenty-first. The scientific gaze of archaeology is informed, I suggest, by the manifold ways in which the Egyptian queen has been imagined, appropriated, and reinvented within the visual culture and historical debates of the last one hundred years.

Revivifying Ancient Femininity

Long before archaeologists began using satellites, filmmakers, artists, novelists, and historians constructed Cleopatra as a decadent spectacle.[36] Arguably one of the most represented female figures in Western history, Cleopatra has been "passion's plaything, sultry queen, a woman so beautiful that she turned the very air around her sick with desire, a tragic figure whose bared bosom made an asp gasp when she died for love."[37] Her face and body have graced objects such as coins and statues. Her beauty has been described by famous writers such as Plutarch and Shakespeare and has been depicted by filmmakers from Georges Méliès to Cecil B. DeMille and Joseph Mankiewicz.

American and European feature films have portrayed the luxurious yet tragic life of the Egyptian queen, and many have used the cinematic apparatus to construct Cleopatra as a sexual and historical spectacle. The lavish mise-en-scène of DeMille's 1934 film, *Cleopatra*, for instance, diverted the

spectator's attention away from the political authority of the ancient heroine (played by Claudette Colbert) and toward the star's sensuous costumes
and DeMille's elaborate sets. Colbert, *Variety* claimed, played Cleopatra as
"a cross between a lady of the evening and a rough soubrette in a country
melodrama."[38] Colbert's lavish and revealing gowns so enchanted female
fans that an entire fashion trend emerged, galvanizing women to recraft
themselves as the queen of ancient Egypt. After seeing the film, women
raced to beauty shops requesting "Cleopatra bangs," and department stores
began carrying expensive Cleopatra gowns and accessories including per-
fume, soap, and cigarette tie-ins.[39] As Mary Hamer explains, "Casting Colbert in the title role and the sale of goods promoted by the film invited
women in the audience to see and equip themselves as Cleopatra."[40]

DeMille was obsessed with historical accuracy in preproduction for
the film. This obsession, which Sumiko Higashi has termed a conflicting "amalgam between realism and sentimentalism," turned historical
specificity toward the service of spectacle.[41] According to one biographer,
DeMille plunged "into immensely detailed research . . . [and] ensured that
even Cleopatra's hairpins were museum copies, [and that] Egyptian water
clocks and Roman candles [were] perfect replicas of the originals."[42] The director addressed the following questions to his research team: "Let us take
that room of Cleopatra's palace at Alexandria, with Julius Caesar seated in
an Egyptian chair. What models of instruments of war would he be playing with? What sort of room was it and how was it furnished? How was
his hair cut? I want to see it all."[43] DeMille spent weeks trying to confirm details such as the correct pronunciation of Cleopatra's name and the
style of her makeup.[44] Such attention to historical detail led DeMille, famous for his opulent visual style, to present viewers, according to *Variety*,
with "ornate and eye-filling pictures of Roman life and Egyptian licen-
tiousness."[45] Despite the film's attempt to meticulously reconstruct ancient
Egypt, however, it conflated Cleopatra's political authority with her sexuality. As Ella Shohat and Robert Stam suggest, DeMille's *Cleopatra* "projected the 'East' as feminine. . . . The sexually manipulative Cleopatra is
addressed as 'Egypt' and . . . the orient is presented as the scene of carnal
delights." They describe such early cinematic representations as a kind of
"quasi-archaeology" for they "enact a historiographical and anthropological role, writing (in light) the culture of 'Others.'" The film, they continue,

Claudette Colbert played
a sensuous Cleopatra
in the 1934 film by Cecil
B. DeMille.

develops an iconography of ancient artifacts "whose existence and revival depends on the revivifying 'look' and 'reading' of the Westerner."[46] This revivifying look of the West, in other words, plays a role in compelling the ancient into being.

Mario Mattoli's 1953 film *Due Notti Con Cleopatra* (*Two Nights with Cleopatra*) also positioned Cleopatra as a seductress, insisting that the queen's intense sexuality could not be contained in just one body. Sophia Loren starred as both the queen of Egypt and her equally statuesque blonde slave, Nisca. Perhaps predictably, *Due Notti* was more about the heroine's voracious sexual appetite than about her political diplomacy. The queen's sexuality is doubled when Nisca, pretending to be Cleopatra, distracts the lusting male guards so the real queen can escape to find Marc Antony. Loren's costumes barely wrapped her voluptuous curves, and she appears almost nude in one scene. A *Due Notti* publicity photo features Loren wearing only a crown, cupping her supple breasts with her hands and staring coyly into

Sophia Loren starred
as Cleopatra in Mario
Mattoli's 1953 film *Due
Notti Con Cleopatra.*

the camera. Mattoli exploited the celebrity of Sophia Loren (who, film historian Stephen Gundle claims, embodied "a highly subversive idea of un-
abashed female sexuality")[47] to create a promiscuous Cleopatra.[48]

A decade later, in 1963, Twentieth Century–Fox released Joseph Mankiewicz's thirty-million-dollar feature film *Cleopatra*, starring Elizabeth Taylor, Richard Burton, and Rex Harrison, one of the highest-grossing films
of its time.[49] As *Variety* insisted, *Cleopatra* was first and foremost about
"spectacle-making."[50] Its reviewer proclaimed, "It's not only a supercolos-
sal eye-filler (the unprecedented budget shows in the physical opulence
throughout), but it is also a remarkably literate cinematic recreation of an
historic epoch."[51] Not only was the film the longest in Hollywood's history
(at four hours and three minutes), it was a "giant panorama, unequalled
in the splendor of its spectacle scenes and, at the same time, surprisingly
acute in its more personal story."[52]

Much of the press coverage for the 1963 film focused on Taylor's portrayal of Cleopatra. In an interesting twist that pointed out the disjuncture
between Taylor's sexual excess and Cleopatra's political panache, one critic
remarked, "Taylor's sexpot Cleopatra certainly had little in common with
the charming, manipulative queen whom Plutarch describes, the woman
who spoke many languages and captivated everyone with her conversa-
tion."[53] The comment reveals the fundamental contradiction within the
figure of Cleopatra (and the feminine more generally)—that is, the inability
to reconcile her imagined sexual lasciviousness with her political throne.
As Caesar admits to Cleopatra early in the film, "You have a way of mixing politics and passion." This dangerous mixing, however, has long tanta-

lized and confounded the Western imagination.[54] As Hamer explains, the
figure of Cleopatra is so potent because it "locates the notion of a woman's
body and the notion of authority together."[55] The cultural fascination with
Cleopatra can be read as symptomatic of Western society's reluctance to
embrace a politically powerful and highly sexual woman.

While the historical epics of DeMille and Mankiewicz construct the
ancient past through enactment of Cleopatra's sexual decadence, other
twentieth-century visual representations of Cleopatra refuse the realism of
the historical film epoch and comment instead on Western culture's very
fascination with the queen. Andy Warhol's 1962 silkscreen *Liz as Cleopatra*,
for instance, features four reproduced images of Elizabeth Taylor playing
Cleopatra dissolving into one another. Using print techniques, the 31'-×-18'
canvas embeds the same Hollywood image of Cleopatra within a printing
process, emphasizing the reproducibility of her image and turning an ancient figure into a modern pop icon. Another work that foregrounds the
simulation and appropriation of Cleopatra's image was featured in a 1994

Elizabeth Taylor played
the queen of Egypt in
Joseph Mankiewicz's
Hollywood spectacular
Cleopatra in 1963.

traveling art exhibit called "Egyptomania!" which intended to "restage a kind of mythic encounter between the modern Western artist and the ancient Egyptian motif."[56] One section of the exhibit, entitled "Cleopatra, or the Seductions of the Orient," featured a small video screen continuously replaying a scene from Mattoli's *Due Notti Con Cleopatra*. Like Andy Warhol's print, this 1954 film loop heightens Cleopatra's hyperrealism, detaching her from the origin myths of ancient Western civilization, mocking the archaeological search for an authentic Cleopatra, and goading the spectator to recognize the ancient past as mechanically reproduced.

While cinematic displays work to contain Cleopatra's political authority by transforming her into a sexual spectacle and artistic renderings mediate on the West's compulsive fascination with and fetishization of the Egyptian queen, I want to explore how such cultural discourses inform the "scientific" gaze of archaeology. In other words, how does the cultural imagination of Cleopatra as "sexually charged" and "infinitely reproducible" shape the archaeological search for her ruins? Where the tension between science and fiction was manifest in DeMille's obsession with historical accuracy in the production of his 1934 fiction film, archaeologists have been conversely preoccupied with Cleopatra's sexuality in their excavation and mapping of her ruins. While this could be understood as a public relations gimmick to attract attention to the excavation, it is also symptomatic of the semiotic excess that surrounds the figure of Cleopatra and that inevitably informs the archeological excavation of her palace. Because of her important role in the master narrative of Western civilization, Cleopatra could never be fully contained as a sexual spectacle. As Warhol's canvas implies, the cultural reproduction of the ancient queen of Egypt has rendered her a site of proliferating and unresolved meaning. This excess only generates further intrigue and mystery, which, as we will see, appears to be as provocative and alluring to the gaze of the archaeologist as to the fiction filmmaker. In her study of Cleopatra's symbolic power, Mary Hamer suggests the ancient queen is a complicated figure: "The political authority (as an historical leader of Greek civilization) lodged within this spectacle means that she cannot simply be fetishized and contained. . . . The prolific cultural reinvention of Cleopatra prohibits an easy fetishization of her body. . . . At the same time, however, the figure of Cleopatra is always highly sexualized in order to mitigate against the political authority she exudes as a pharaoh and significant political leader in the history of Western civili-

zation."[57] It is precisely the combination of Cleopatra's sexuality *and* her
place in the political history of Western civilization that makes her such a
complex and enigmatic figure.

In part because of this, Cleopatra has been subject to other forms of revisionism as well. In 1996 the Smithsonian's National Art Gallery hosted
an exhibit called "Lost and Found: Edmonia Lewis' Cleopatra," featuring a
restored nineteenth-century sculpture entitled "The Death of Cleopatra,"
made by the first known African and Native American sculptor. The statue
rested first atop the grave of the famous racehorse Cleopatra, near Chicago,
and then remained in a storage yard for decades. During the late 1980s the
chipped and graffiti-covered sculpture underwent a thirty-thousand-dollar
restoration. Lewis, who spent much of her adult life working as a sculptor in Rome, used classical figures to obliquely comment on the plight of
African Americans and their struggle for emancipation. Lewis's sculpture
was first exhibited at the Centennial Expo in Philadelphia in 1876, and according to *The Peoples' Advocate*, an African American weekly, " 'The Death
of Cleopatra' excites more admiration and gathers larger crowds around it
than any other work in the vast collection of Memorial Hall."[58] Although
most artists of the period depicted Cleopatra in a "calm, idealized regal like-
ness," as Stephen May explains, "Lewis created a fundamentally different
image; in a break with the neoclassical canon, which downplayed strong
emotions, she portrayed the queen at the point of death."[59]

Historian David Driskell interprets "The Death of Cleopatra" as Lewis's
attempt to forge a symbolic connection between Egyptian and African cultures. He reads the piece as a kind of self-portrait: "If you look at the portrait
of Edmonia, you see clothing lavishly placed all over her body, very much
like she's placed the clothing over Cleopatra's body." He continues: "I think
she was trying to show how the personality of Cleopatra was so strong that
she would triumph even in death. She would command attention even in
death."[60] Notably, the only known sculpture of Cleopatra that portrays her
as dead was created by an African American artist who engaged in an interrogation of the circumstances of Cleopatra's life, death, and racial identity. In the decades since Lewis's sculpture Cleopatra's identity has become
a rallying point around which Afrocentrists have reconceived the relationship between ancient African history and the Western world.

In the past two decades a number of revisionist Afrocentric histories
of the classical period have emerged. One of the most renowned has been

Martin Bernal's two-volume *Black Athena: The Afroasiatic Roots of Classical Civilization*. In it Bernal critiques what he calls the "Aryan Model" of Western classical historiography, which has ignored the Egyptian and Phoenician influences on Greek civilization.[61] If we accept his revisionist history of the classical period (which he calls the "Ancient Model"), Bernal explains, "it will be necessary not only to rethink the fundamental bases of 'Western Civilization,' but also to recognize the penetration of racism and 'continental chauvinism' into all our historiography."[62] Bernal insists classical Western historians have resisted the idea that Greek civilization, "which was seen not merely as the epitome of Europe but also as its pure childhood," is the "result of the mixture of native Europeans and colonizing Africans and Semites."[63] Calling for a reinterpretation of Western Greek origin myths, Bernal argues that Alexandria was a hybridized, multiracial city of various cultural influences.

Both before and after Bernal's book, Afrocentrist scholars claimed that Socrates and Cleopatra were of African descent and that Greek philosophy was stolen from Egypt. Many classical historians were outraged and responded aggressively. Mary Lefkowitz, for instance, published a book called *Not Out of Africa: How Afrocentrism Became an Excuse to Teach Myth as History* as an incensed retort to such propositions. In it she refers to Afrocentric scholarship as "pseudohistory" and emphasizes the importance of teaching future generations about the "real ancient Egypt" and the "real ancient Africa."[64] In a chapter called "Was Cleopatra Black?" Lefkowitz ridicules Afrocentric historians and cultural critics for raising questions about Cleopatra's racial identity and chides them for failing to produce primary evidence.[65] It is not enough, Lefkowitz claims, to cite Shakespeare passages that refer to the Egyptian queen as "tawny" or "black." After systematically refuting claims about the possibility of Cleopatra's African ancestry, she argues for an ancient history based on "acceptable proof."[66] What is significant here is not so much the empirical question of biological heritage but the ongoing investment of scholars (on both sides of the argument) in what is essentially a battle over cultural heritage, a battle that seeks to find in the ancient past the political, cultural, and racial antecedents for contemporary society. Put another way, the narrow "scientific domain" of studying Cleopatra's world has mushroomed into a full-fledged struggle over the "archaeological territory" on which Western culture rests.

Tamara Dobson played a globetrotting
CIA agent in the 1973 blaxploitation
film *Cleopatra Jones*.

In the context of these intellectual debates Cleopatra is positioned
between continents, between the North and the South, and between Eurocentric and Afrocentric epistemologies. Such debates have directed attention to this ancient icon and transformed her into a site of popular reclamation and redefinition, particularly among African Americans. Consider, for
instance, the resurging popularity of 1970s blaxploitation films featuring
Cleopatra Jones (an African American CIA agent who fights the world drug
trade in an effort to bring social reforms to local black communities),[67] the
African American female rap trio Cleopatra, the character Foxy Cleopatra
played by Beyonce Knowles, or Raven Symone's (child actress of *The Cosby
Show*) decision to name her big black dog Cleopatra "because she's a beautiful black queen."[68] A book about the life of Josephine Baker nicknames the
popular African American singer "Jazz Cleopatra."[69] And, African American writer Mike Holt proclaims that one reason the black press is so important is because "they [whites] print ads showing Elizabeth Taylor portraying
Cleopatra. In truth, the queen of the Nile looked more like Halle Berry."[70]
Such black feminine pop icons, like Edmonia Lewis's sculpture, deploy the
symbolic meanings of the ancient Cleopatra to form powerful identities in
the present. It is the uncertainty and ambiguity of Cleopatra's racial heritage, produced in part by the reproduction of her image, that enables such
a practice.

Cultural discourses on Cleopatra—whether in films, art exhibits, his-

torians' debates, or African American press and popular culture—engage
the figure of Cleopatra with a range of social and political positions in the
present. As Hamer contends, "The figure of Cleopatra . . . [has been] invoked to bring into focus the challenge, if not the threat, implied by rethinking Western culture from an Afro-American perspective."[71] The spectacle
of Cleopatra—her position as an object of the gaze—thus also triggers an
interrogation of her place in history. The more deeply we inquire into Cleopatra's past, in search of the key to her political power, the more questions
emerge.

Rather than resolve those questions, the application of the satellite gaze
to Cleopatra's watery grave has only further stirred them. As we will see
in the final section of this chapter, the archaeological investigation into
Cleopatra's life merges two forms of knowledge that are commonly treated
as distinct: voyeurism and scientific objectivity. In one sense satellite archaeology functions as a twentieth-century technologization of what Anne
McClintock calls the "porno-tropics for the European imagination." During the colonial period, McClintock explains, Enlightenment metaphysics
often "presented knowledge as a relation of power between two gendered
spaces, articulated by a journey and a technology of conversion: the male
penetration and exposure of a veiled, female interior; and the aggressive
conversion of its 'secrets' into a visible, male science of the surface. . . . In
these fantasies, the world is feminized and spatially spread for male exploration, then reassembled and deployed in the interests of massive imperial
power."[72] As archaeologists swim through the feminized space of Cleopatra's underwater palace, they endeavor to bring its secrets to the surface.
The satellite images and GPS maps that represent the site of her excavation, in all their claims to accuracy and objectivity, must borrow from the
same systems of representation they attempt to replace. Satellite images,
in other words, are not detached from other cultural forms; rather, as I have
suggested, they must rely on them to signify anything other than their own
remoteness and abstraction.

Archaeological Voyeurism

The excavation of Cleopatra's palace began in 1993, when French marine
archaeologists Jean Yves Empereur and Frank Goddio first launched the
campaign to excavate the underwater ruins.[73] A year earlier Goddio had

A diver uses a GPS device to mark the location of an underwater artifact.

conducted a sonar examination of the seabed in Alexandria's eastern harbor, a protected military zone, but interference from the noisy city made it inaccurate. He then successfully turned to satellite images of the region, together with ancient texts, to locate the ruins more precisely.[74] In April 1996 Goddio led a team of sixteen professional divers and twenty archaeologists and computer experts in a campaign to map the quarters of Cleopatra and Marc Antony.

By November 1996 Goddio's team had made more than thirty-five hundred dives and used underwater camcorders and handheld GPS receivers to reportedly "pinpoint exactly where Cleopatra and Antony once walked."[75] When the GPS receivers were mounted on rubber dinghies and combined with a sonar device, they gave an almost exact reading of the contours of the seabed. Divers also swam around the ruins with GPS receivers gathering precise spatial data, and a team of computer experts transformed this data into a map of the two-thousand-year-old city.[76] The map specifies the locations of the enormous dikes that once enclosed the ancient Egyptian city, the foundation of Cleopatra's palace, the lighthouse that was one of the Seven Wonders of the World, the remains of the royal galleon *Timonium*, and the Temple of Poseidon. As the London *Times* proclaimed, "Barely 20 feet beneath the placid surface of the Eastern harbor of Alexandria lies a treasure trove belonging to two of the most evocative names of ancient history, Antony and Cleopatra."[77]

In November 1996 Frank Goddio, wearing a T-shirt that read "Cleopatra" in hieroglyphics, escorted reporters and camera crews on a tour of the Alexandria site in the glass-bottom ship *Oceanic*. He pointed out dozens of artifacts, including Greco-Roman busts, pharaonic obelisks, sphinx figures, and architectural ruins.[78] Goddio insisted that "the most important find of all [was] the palace of Cleopatra on what was once the island of Antirrhodos," a palace described by archaeologist Dr. El Faharani as "an unbelievably lavish palace painted in many colours with multi-layered gardens," not unlike Mankiewicz's sets.[79] During their tour Goddio invited reporters to look through an underwater camera at the remains of Cleopatra's palace.[80] Goddio explained how electrifying it was to actually touch the artifacts below, proclaiming, "It was a fantastic feeling diving on the remains of the city. To think that when I touched a statue or sphinx, that Cleopatra herself might have done the same."[81]

After touring Cleopatra's underwater palace, reporters generated many stories about the excavation. Much of the coverage reinforced cultural discourses on Cleopatra's sexuality, describing the archaeological site as a sensuous wonderland. Newspaper accounts underscored the sexuality of the artifacts themselves through such headlines as "Cleopatra's Playground Revealed," "Divers Discover Cleopatra's Alexandrian Trysting Place," and "City Beneath the Sea Surrenders Its Secrets."[82] *Newsday* referred to Alexandria as "the glittering city where Queen Cleopatra VII cavorted with her Roman lover Marc Antony nearly 2,000 years ago."[83] And another described the site as "the royal compound where Egypt's Queen Cleopatra adorned herself to seduce the mightiest men of Rome twenty centuries ago."[84] As "the most renowned femme-fatale of the Graeco-Roman era," according to *Deutsch Press-Agentur*, Cleopatra "began her romantic engagements with the Romans by seducing Julius Caesar when he came to Egypt in 48 BC."[85] Finally, *Time* indicated, "Scientists are most intrigued by the . . . toppled pillars and cobbled pavement . . . [that] may mark the spot where Cleopatra ruled between 51 BC and 30 BC and fell in love with Mark Antony."[86]

Such accounts positioned Cleopatra's sexuality itself as the object of archaeological investigation. This voyeuristic fascination was also manifest

Reporters and archae-
ologists watch as torsos
and busts are lifted from
the underwater excava-
tion site in Alexandria.

in the way that artifacts were lifted from the sea and revealed to the world by
the media. In a cnn report, for example, the narrator excitedly describes the
excavation's findings, and as a large female bust emerges from the water,
she states, "Archaeologists lifted the torso of a pharaonic goddess into the
sunshine, shedding new light on long buried treasures that, until now,
could only be imagined."[87] Similarly, a pbs documentary titled *Treasures
of the Sunken City* features a scene in which crowds of reporters and on-
lookers watch enormous torsos and sphinxes emerging from the depths
as if the ancient queen of Egypt herself were resurfacing.[88] In a scheme
that combines cinematic spectacle and archaeological discovery, Goddio
plans to transform the site into an underwater museum that can be viewed
through glass-bottom boats or transparent walkways. As he explains, "We
want people to be able to see the site as it is."[89] This plan would turn Cleopatra's "trysting grounds" into an accessible domain that visitors could pass
through to scrutinize the queen's private chambers.

A Discovery Channel documentary about the excavation, *Cleopatra's Palace: In Search of a Legend*, explicitly links the satellite to archaeological voyeurism. The film's narrative is structured around the gaze of Frank Goddio,
but in an interesting reversal the opening shot features an extreme close-up
of a woman's eyes (implicitly Cleopatra's). As Goddio explains, "I've always
been dreaming of coming to this place because of the mystery of this site."
It is her eyes, the sequence suggests, that he ultimately dreams of finding.
The narrator continues: "A former financial adviser, he [Goddio] abandoned
his successful career to pursue Cleopatra full-time, and this is where he

hopes to find her." Next we see a shot of Goddio diving underwater as his flashlight lights up the muddy surface of the ruins. Later, on his boat, he reveals a ring and hairpin he discovered during a dive as if he is getting closer to finding the woman that once wore them. While underwater, we are told, "he is touched by the nearness of Cleopatra's spirit."

As the narration of *Cleopatra's Palace* positions Goddio as a voyeur, it also incorporates a series of orbital views. One sequence connects the satellite's gaze to the history of cultural fascination with Cleopatra as the narrator explains that by using remote sensing and GPS, the archaeologist "is finally able to bring the lost world into view. His map offers entry into the queen's domain, a palace where Cleopatra seduced Julius Caesar." Then GPS maps of the excavation appear, and they are replaced with orbital views of Alexandria. The satellite perspectives are intercut several times with fiction-film clips and dramatic enactments showing Cleopatra seducing her lovers, Julius Caesar and Marc Antony. Here the archaeological use of satellites is constructed as *part of* Cleopatra's legacy as a cultural spectacle rather than as the resolution of it. The orbital gaze "finds" fictionalized accounts of Cleopatra's escapades rather than simply pinpointing an excavation site beneath the earth's surface. This alternation is key, then, for it posits the orbital gaze and cultural imaginary in relation to one another rather than as discrete entities. Further, these sequences complicate notions of scientific objectivity and distant observation because they literalize the terms *remote sensing* and *global positioning* by visualizing macro-micro, worldly-bodily, and cosmos-self dialectics.

The alternation of satellite images and dramatic enactment also uncovers the fundamental impulse of archaeology—that is, the rationalization of desire for visual and tactile sensation of the ancient world. Within the film's narrative of archaeological discovery the satellite image functions as a master shot motivating a desire not just for proximity but tactility as well. For it is the possibility of physical contact with the past that is the key lure of archaeology in general, and the unique appeal of Cleopatra makes such a proposition even more compelling. As Goddio himself has confessed numerous times, one of the pleasures of diving at the site is being able to touch the very statues that Cleopatra herself once touched. In another documentary Goddio is featured kissing one of the pillars of Cleopatra's palace as if he were kissing the ancient queen.

What archaeologists are really making contact with are not Cleopatra's lips but the GPS receivers and cameras used to survey the ruins. The desire for physical contact with the ancient world is gratified, in other words, through use of handheld media technologies that are imagined as augmenting pick and shovel. Cameras, GPS receivers, and sensors are carried and maneuvered to generate interactive maps of ruins for others to tour and touch. Divers' underwater jaunts culminated in a Web interface that enables users to navigate the ruins remotely. On the *Treasures of the Sunken City* Web site an interactive GPS map indicates the locations of 2,110 artifacts represented as red blocks scattered across the seafloor of Qait Bay.[90] By clicking on the red blocks, the user can access close-up underwater photographs of the artifacts. For instance, artifact number 1,020 is a statue base; number 1,008 is the Sphinx of Psammateichus; and number 1,001 is a colossal masculine statue; numbers 1,003 and 1,027 are massive slabs thought to be parts of buildings.

The GPS map is not just an abstract rendering of locational data, however. It is infused with the particular perspectives of divers who swam among the shambles of ancient Alexandria with underwater cameras.[91] In this sense the GPS interface shares a structural resemblance to remote television production as close-up views "on location" are embedded within a broader mapping of the site. The remote sensing of Cleopatra involves a transition from cinematic to televisual and digital modes of cultural production since discourses surrounding the excavation transform Cleopatra's spectacle into a remotely accessible and navigable digital domain that can be energized with the touch of a keyboard or a mouse.

While Web interfaces construct an experience of tactile contact with ancient Alexandria, discourses surrounding the excavation point the televisual gaze toward Cleopatra's body, which, of course, no longer exists, but the desire for it is displaced onto statues, busts, and other ruins. As Carolyn Marvin suggests, "The body is a convenient touchstone by which to gauge, explore, and interpret the unfamiliar, an essential formation-gathering probe we never quite give up, no matter how sophisticated the supplemental modes available to us."[92] In this case Cleopatra's body becomes an empty signifier that functions to deflect attention back on the gaze itself and interrogate how it is constituted and operationalized.

Indeed, the documentary *Cleopatra's Palace* is as much about the ubiq-

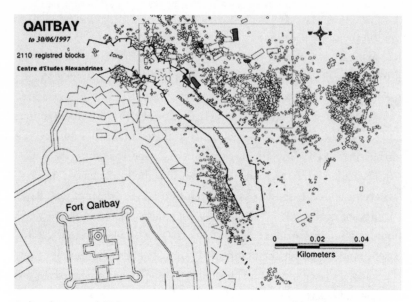

Archaeologists used GPS receivers to generate interactive maps of the excavation site.

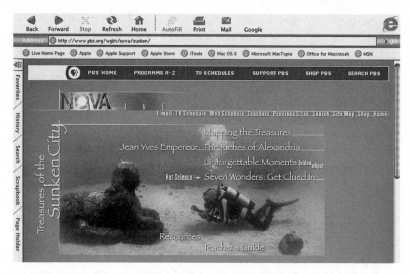

The interface for the *Treasures of the Sunken City* project is designed in a way that transposes practices of Web navigation and underwater excavation.

uity of the televisual gaze as it is about an archaeological excavation. Orbital and underwater views are themselves fetishized as spectacles throughout the film. In other words, we are looking at a series of technologized looks as much as we are "discovering" Cleopatra's ruins. Imaging technologies ranging from satellite sensors, to computer compositing, to underwater cinematography are prominently and repetitively displayed as distinguishing features of this excavation. One sequence of *Cleopatra's Palace* employs a digital zoom to position the viewer as being able to "fly" continuously from orbit and then dive underwater to "swim" the ruins. Like *Our World* such sequences reinforce discourses of global presence, interpellating the viewer as able to be anywhere and everywhere at once with an illusion of fluid and seamless motion. While *Our World* revels in the global now, however, *Cleopatra's Palace* uses this liquid visuality to forge continuities between ancient and modern, science and culture, ocean and outer spaces, spaces that Mette Bryld and Nina Lykke call the "wild elsewheres" of modern science. As they suggest in their book *Cosmodolphins*, "Just as imaginary world maps of contemporary culture are marked by a 'high frontier,' so they are also demarcated by a 'deep frontier.'"[93] In *Cleopatra's Palace* the televisual negotiates these fields, bringing them within range of the same apparatus and enabling archaeologists and TV viewers to ride high and deep frontiers in search of antiquity.

Conclusion

In this chapter I have explored the remote sensing of Cleopatra through the convergence of satellite, television, and computer technologies. I have suggested that such convergence generates and extends particular televisual epistemologies—ways of seeing and knowing the world from a distance that are structured by and contingent on specific sociohistorical, cultural, and technological conditions. When televisual epistemologies are mobilized and reconfigured within disciplines such as archaeology, they can unsettle or disturb their scientific rational imperatives. But archaeology, as I have also emphasized, is not just a technique of preserving the ancient past. It is also a critical practice of analyzing the emergence of discursive objects and their relation to regimes of knowledge, power, and truth. Archaeology is, as Foucault puts it, "to write a history of discursive objects that does not plunge them into the common depth of a primal soil, but deploys the nexus of regularities that govern their dispersion."[94]

I have treated the remote sensing of Cleopatra as a discursive object and attempted to reveal the "nexus of regularities"—the technologies, ideological systems, origin myths—through which it has been actualized. The television documentary *Cleopatra's Palace* best encapsulates some of the broader themes I have emphasized. It reveals, on the one hand, the ways in which archaeology might be construed as a televisual practice both by virtue of its participation in broadcasts about the excavation and in its own reliance on video camcorders, satellite images, and GPS maps to produce views of the ancient past. (There is only one archaeologist—Frank Goddio—prominently featured, but there are many imaging technologies shown.) On the other hand, *Cleopatra's Palace* challenges archaeology's claims to scientific objectivity by interlacing its gaze with fictional discourses on Cleopatra. *Cleopatra's Palace* combines satellite, digital, video, and film sequences so that they blend into one another undetectably, making distinctions between scientific observation and fictional representation nearly indiscernible. In this sense the structure of the documentary resonates with a broader goal of this chapter—that is, to challenge the assumption that remote sensing and GPS simply augment the archaeological gaze, making excavation more scientifically efficient, precise, and objective. As I have tried to suggest, archaeological uses of satellites are themselves informed and shaped by the discursive objects that they seek.

In this case satellite television's modality of scientific observation alternates with pleasurable acts of seeing and knowing from a distance—or voyeurism, which, of course, is a defining structure of commercial entertainment. Since the discourses surrounding this excavation focus so heavily on Cleopatra's sexuality, they suggest a reconfiguration of voyeurism that is bound up with the particularities of technological convergence. In this instance the media discourse on Cleopatra's sexuality shifts from one of cinematic spectacle "to be looked at" to a televised excavation "to be watched" to a digital interface "to be navigated." Because Cleopatra is a discursive object with such leaky seams, multiple systems of technologized vision and voyeurism must be harnessed to regulate her historical significance. And the more fuzzy the logics between science and fiction, the orbital and oceanic, the ancient and modern, the East and West in Cleopatra's representation, the more pronounced the compulsion to fix her position.

Satellite television can be understood as a practice of archaeology to the extent that it is used to conjure and revivify discursive objects that either no longer exist or can only be detected from a distance that the human eye cannot occupy. Rather than proclaiming to see round the world in an instant, archaeologists can claim to spot long-gone civilizations in one orbital view. And while this view may be delimited by edges of the frame, the televisual gaze that generates it extends back in time. The collision of archaeology and television thus also results in practices of diachronic omniscience, which attempt to stretch distant vision not only across global space but through global time as well. However, the further the televisual gaze peers back in time, the more forcefully it collapses into the realm of the senses. For distant vision needs to be embodied. That is why the term *remote sensing* is so apt for the televisual, because it implies that distant vision ultimately necessitates and is contingent on various visceral experiences, libidinal investments, and sensuous engagements that could never be reduced to the visual alone. Understanding the televisual as a practice of remote sensing thus complicates paradigms that attempt to isolate and privilege vision as the ultimate sense, as the one of greatest accuracy, knowledge, and truth, and insists on the significance of other senses, whether hearing, touch, or taste, in interpreting the world.

In his writings on the digital image, William J. Mitchell describes the reams of satellite images that have accumulated since 1962 as the earth's "ceaselessly shed skins."[95] This phrase is provocative because it links remote sensing with embodied processes of change and renewal, and it enlivens an understanding of the earth's surface as a text. Indeed, satellite images intextuate atmospheric conditions surrounding, geological layers beneath, and unfolding events upon the earth's surface. If we accept the earthly skins metaphor for satellite images, perhaps we will be more inclined to treat them as fields of embodiment and less apt to overlook the lifeworlds submerged within them.

One of the twenty-seven dishes at the Very Large
Array in Socorro, New Mexico. Photo by Matthew L.
Abbondanzio. Courtesy of NRAO/AUI.

Satellite Panoramas

ASTRONOMICAL OBSERVATION AND REMOTE CONTROL

When the televisual is not fixed as a technology of broadcasting, its meanings can be expanded and understood in relation to practices such as
military intelligence, archaeology, and astronomy—practices that employ
technologized vision to gaze on and produce knowledge about the earth's
surface, the ancient past, and the cosmos. The televisual is inherent to these
fields of inquiry because the knowledges they generate are based on modes
of distant observation. This chapter considers astronomical observation as
a televisual practice. In it I explore how images generated by, discourses
surrounding, and texts symptomatic of the Hubble Space Telescope work
to extend and alter the meanings of the televisual. I analyze media sites
such as broadcast news, print journalism, IMAX documentary, and Hollywood cinema to establish, as in chapters 3 and 4, the range of intermediale
practices through which the televisual is derived. Much more than a system
of commercial entertainment or public broadcasting, the televisual is an
epistemological system that can be operationalized across different media
formats and that has alternating modalities. Televisual epistemologies are
the specific knowledges and knowledge practices that form around technologized acts of distant observation, and they migrate far beyond the TV
set and the living room.

On the eve of the Hubble Space Telescope's launch in April 1990, NBC
TV's John Chancellor delivered an editorial to American television viewers about NASA's latest project, proclaiming, "Earthlings! A New Age is
dawning! Galileo first pointed a telescope 400 years ago and since then
we've been looking and wondering how the stars got there and how the
universe began." Celebrating Hubble's unprecedented vision, Chancellor
excitedly declared that pre-Hubble astronomy was like "looking at birds

from the bottom of a swimming pool. What we saw was wavy and murky. Next week . . . the human race will climb out of the swimming pool and begin looking at pictures of the universe that are as clear and sharp as an etching. . . . Nothing in space exploration comes even close to the potential of the Hubble Space Telescope." What Chancellor's rhapsodic commentary suppressed were the heated debates surrounding Hubble's development, documented by Eric Chaisson in his book *Hubble Wars*, which not only questioned Hubble's two-billion-dollar price tag but challenged its scientific utility as well.[1] Such skepticism was only exacerbated by the fact that once Hubble reached orbit, astronomers discovered it had a defective mirror. On May 20, 1990, in much less celebratory terms, NBC broadcast Hubble's first image—a blurry black-and-white photo of an ancient star cluster that looked like an abstract and colorless blob. This early technical failure disappointed European Space Agency (ESA) and NASA officials who had developed Hubble as part of a cooperative agreement to operate a long-term space observatory for the benefit of the international astronomical community, not unlike broadcasters decades earlier who collaborated to use communications satellites to craft a "global village."[2]

Determined to protect their financial investments and scientific reputations, the ESA and NASA orchestrated a delicate repair to Hubble's reflector in December 1993. CNN carried live footage of astronauts as they performed Hubble's spectacular "eye adjustment," and a month later fresh Hubble images began to circulate in the news media. At an overflowing news conference on January 14, 1994, U.S. and European officials showed "before" and "after" Hubble test images revealing a dramatic improvement following the repair of its mirror. Senator Barbara Mikulski declared, "The trouble with Hubble is over!" The *Washington Post* reported that Hubble's "new focus" replaced what once appeared to be "fuzz balls" with pictures of deep space that allowed us to "almost see forever."[3] Space scientist James Crocker boasted that "if the Hubble [were located] in Washington, it could detect fireflies in Tokyo."[4] And astronomer Terence Dickinson insisted that Hubble's fixed vision was so clear it would "rewrite the textbooks."[5]

Hubble functions like a remote-sensing satellite, but its gaze is directed toward deep space rather than Earth's surface. Since Hubble moves in a low-Earth orbit, 332 miles above Earth's turbulent atmosphere, it is able to deliver remarkably sharp images of deep space phenomena.[6] The satel-

lite generates images by sensing light and/or heat in outer space, storing
it as data, and transmitting that data to receivers on Earth. Once the data
is received by Earth stations, it is archived and can be processed and displayed on computer monitors or reproduced in newspapers, magazines,
videos, Web sites, or films. When Hubble images move beyond the ground
station and into the mediascape, they become satellite panoramas. These
panoramas are not tranquil vistas of outer space; rather, they are contested
discursive terrains that are activated as the conventions of astronomical
observation, digital imaging, broadcasting, and filmmaking are combined
and brought to bear in their production, circulation, and interpretation.

In this chapter's first half I explore how news media have treated Hubble
images as live media events and cosmic sonograms, focusing on the Shoemaker-Levy 9 comet's collision with Jupiter and processes of star formation respectively. In both instances Hubble images of space phenomena
are figured as part of world history and human bodies as planetary events
and stellar matter are imbued with earthly significance. In the latter half of
the chapter I explain how the films *Cosmic Voyage, Contact,* and *The Arrival*
invoke Hubble images and practices of astronomical observation, appropriating its capacity for distant vision, only to open a space in which to assert a
discourse of remote control. As we will see, this discourse is articulated in
the form of evolutionary logics in *Cosmic Voyage,* in official science's refusal
to reconcile the emotive experience of an astronomer in *Contact,* and in the
containment of alien signals transmitted across the U.S.-Mexico border in
The Arrival.

Indeed, throughout this chapter I suggest that embedded within Hubble's capacity for distant vision is a discursive strategy of remote control.
This strategy further elaborates what scholars have identified as mass culture's domestication of outer space. Both Lynn Spigel and Jodi Dean, for
instance, have explored how popular media have historically domesticated
the strangeness of outer space by grounding it within cultures of everyday
life, whether the fantastic family sitcoms of the 1960s or user-friendly Web
sites of the 1990s.[7] Dean insists that space-age fantasies of outer space mobility have shifted in the digital era. As she explains, "Previously space was
linked with the agency of the astronaut; it now connotes the passivity of the
audience who witnessed the conquest of space on television. . . . The cultural stress had been on escaping the confines of earth; now it's on finding

ways to stay home."[8] It is this desire to explore outer space only to remain home, grounded on Earth, that triggers the central questions of this chapter. For this desire is also symptomatic of a refusal to embrace the potentials for difference, unknowingness, and uncertainty that constitute and give significance to outer space.

By considering Hubble's relation to a range of media contexts, I explore how its astronomical observations not only alter and extend the televisual but also, like the remote-sensing practices discussed in chapters 3 and 4, summon and reinforce a broader set of Western Eurocentric assumptions. The discursive strategy of remote control, I argue, is a key mechanism for negotiating the extreme temporalities, distances, and scales that define Hubble's satellite panoramas. The farther Hubble gazes into space, the more tightly its otherworldly images are anchored to human embodiment and world history. Thus, as astronomers proclaim that Hubble is revealing phenomena never known to exist before, this most remote and incomprehensible matter, it seems, can only be understood as an extension of Western knowledge, history, and scientific progress. Just as archaeologists' gaze backward in time compelled a search for the materiality of Cleopatra, the deeper Hubble gazes into space, the more elastic the discourses surrounding its vision become. For each remote finding snaps us back forcefully to our world. What Hubble ultimately reveals, then, are the limits of the Western imaginary and the necessity of understanding its televisual gaze as part of an epistemologic system as opposed to one of naked discovery.

A Media Event of Galactic Proportions

On February 8, 1994, two months after Hubble's delicate repair, NBC broadcast a dramatic Hubble image of the Shoemaker-Levy 9 comet plummeting through space. The broadcast was symbolic because it verified Hubble's improved vision and foreshadowed its involvement in a media event of galactic proportions. A few months later, heading "the largest telescope armada ever assembled in the history of astronomy to observe a single event," Hubble captured images of that same comet as it slammed into Jupiter.[9] Astronomers tracking Shoemaker-Levy 9 with ground telescopes had predicted that the comet's fragments would hit Jupiter in late July 1994, and U.S. news media responded by building anticipation of the impending col-

Hubble images revealed the Shoe-maker-Levy 9 comet's collision with Jupiter in July 1994.

lision. On May 20, 1994, *Time* plastered Shoemaker-Levy across its cover. News reports proclaimed the comet, which earned such nicknames as the "cosmic dust bunny," the "gargantuan fireball," and the "doomsday rock," would represent the first time ever that humans witnessed the kind of disaster that first gave shape to the earth and human civilization, and insisted it was a "must see."[10] An article titled "Planetary Punishment: Comet to Spank Jupiter's Behind" told readers they could "witness for the first time ever the kind of event that put craters on the moon and may have killed the dinosaurs."[11] CNN aired a NASA computer simulation of the collision predicting that "the event . . . [would] be the biggest explosion ever seen by human beings."[12] Placing the catastrophe within the context of modern world history, *Agence France Presse* reported that if the comet had struck the earth, its impact would equal fifteen million Hiroshima bombs.[13] Another report compared the comet's crash to chunks the size of Mt. Fuji falling out of the sky at 140,000 miles an hour.[14] Eager to witness the event, amateur astronomers flocked to the desert of New Mexico to catch it through their own telescopes.[15] The Nature Company, in fact, tripled its sale of telescopes that year.[16]

With astronomers and journalists around the globe transfixed, the cosmic collision, news media insisted, would be a "live media event." In the days before the crash U.S. television networks monitored the comet's trajectory and predicted where it would make contact. CNN promised viewers to bring "pictures as we get them from NASA, the Hubble Space Telescope and other astronomers with private telescopes."[17] NBC announced that Hubble would take hundreds of pictures of the "biggest cosmic event in recorded history," and the network showed pictures of professional and amateur astronomers working side by side. When fragments of Shoemaker-Levy 9 finally hit Jupiter at 4:00 p.m. EDT on July 16, 1994, Hubble's images were beamed around the earth almost instantly. On the third day of the crash, an NBC anchor reported that "astronomers have never seen anything like it" and explained that the comet's G (or seventh) fragment had "slammed into Jupiter at 130,000 miles an hour, [with] four hundred times the power of the entire world's nuclear arsenal."

Such references to nuclear annihilation and global cataclysm were symptomatic of a more general premillennial angst manifest in events ranging from the Heaven's Gate mass suicide to the Columbine massacre. As Susan Buck-Morss insisted in her mid-1990s book *Ground Control*, "The air is full of rumours and announcements of various terminations, of the end of humanity, of the end of history, of the end of the planet."[18] Hubble images of Shoemaker-Levy 9's collision became apt icons for an imagined end to civilization at the end of the twentieth century, but the very fact that earthlings could witness the event on television meant that while Jupiter may have been hard hit, Earth was still intact.[19]

The coverage of the collision fit neatly within the rubric of "catastrophe television," which is characterized by on-site observation of disasters, play-by-play dissection of horrific events, and repeated computer simulations.[20] Since news teams could not literally go on-site, they relied on Hubble images and digital animation to dramatize the event. For instance, one NASA computer simulation of the comet's collision replayed on the major TV networks begins with an image of Jupiter's "impact scar," which, the announcer explains, is "10,000 miles wide and 4,500 miles long." As a picture of Earth pops into the frame, the announcer tells viewers that Jupiter's scar spans an area larger than our entire planet. The image of Earth shrinks and, like a comet itself, plummets into Jupiter's newest impact crater, implying the event was so massive it would have eviscerated the earth in a flash.

The sequence ultimately suggests the horrifying scale of the event while highlighting our distant position from it and hence our capacity to simulate and analyze it. As Patricia Mellencamp explains, "Catastrophe coverage can function to ensure our feelings of well-being and good fortune. . . . The assurance for the viewer is that it is happening, but elsewhere; or, it has already happened and is now historical, over. The tantalizing threat, the true danger of catastrophe is the here and now; if it were happening to us, we wouldn't be watching television."[21] Indeed, commercial television could only ever represent something comparable to nuclear annihilation "live" if it were occurring elsewhere. This practice of presenting live coverage of catastrophic events happening elsewhere has become one of the dominant television news conventions of our time. News media integrated Hubble images of Jupiter's disaster as if fed from any other remote camera, whether deployed to cover an earthquake in Turkey, the monsoons in India, or war in Iraq. Since Hubble functions, in effect, as TV news networks' most remote camera, its views work to further extend and dramatize the medium's audacious promise of unlimited access to unfolding events, constructing televisual presence not just as global, as discussed in chapter 1, but as cosmic as well. In this way the collision coverage expanded the "as the earth spins" discourse of early live global television, suggesting that just as the earth spins on its axis, network television can be deployed to reveal anything anywhere not just *upon* the earth but *beyond* it.[22]

Also at issue here are the ways in which Hubble images become part of televisual epistemologies. Viewers' knowledge of Jupiter's catastrophe is derived through the process of inserting Hubble's rendered image data within the conventions of live network television. Since Hubble images are digital, they lack physical referents, but they were presented *as if* live television views of a galactic event.[23] However, not only were we seeing an event on another planet, but it had already happened billions of light years ago! Still, the conventions of live television were used to frame the comet's collision as if it were occurring while we were watching.

The treatment of Hubble images as live event coverage also raises interesting questions about the relations between the televisual, the digital, and the historic. Hubble's convergence with network television — that is, the inscription of its digital renderings of events that occurred light years ago within the frame of an instantly transmitted analogue signal — denaturalizes "liveness" as a unified simultaneity and exposes it as an assemblage of multiple media and temporalities. As Jane Feuer reminds us, live television is "best described as a collage of film, video and 'live,' all interwoven into a complex and altered time scheme."[24] As an intermediale form, satellite television becomes a series of alternations between analogue and digital formats, present and past tenses, earthly and cosmic scales.

Whereas *Our World* used satellite television technology to contrive a historic "globe-encircling-now," Hubble images of the Shoemaker-Levy 9 collision were used to turn another planet's disaster into our world's televised history. As I have argued throughout this book, the production of historical knowledge is increasingly bound to and regulated by particular uses of satellite television, whether the international transmission of distant happenings, the remote sensing of the earth's surface, or the astronomical observation of cosmic events. What the situation of the Shoemaker-Levy 9 collision reveals is the high level of authority we invest in televisual technologies to determine the spatiotemporal bounds of the historic and to regulate our knowledge of it. As satellite television pulls cosmic events into view, the televisual is rearticulated as a ubiquitous apparatus that can withstand the ultimate disaster, accommodate multiple media formats, traverse unfathomable distances, and deliver spectacular history.

Cosmic Sonograms

Hubble images have been assimilated within live media events in ways that alter the televisual and the historic, but they have also been used to represent extremely distant matter as if it were part of us. When the Hubble Deep Field image revealed a horde of other galaxies in 1996, astronomer Robert Williams admitted, "As the images have come up on our screens, we have not been able to keep from wondering if we might somehow be seeing our own origins in all of this."[25] Since Hubble's gaze is directed back in time, its images are said to chronicle the birth and evolution of the earth and the cosmos. Astronomer Carols Frenk likens Hubble's array of images "to having a snapshot of different individuals at selected stages of development—say a baby, an infant, a teenager, and an old person—and then trying to piece together a theory of human growth."[26] Another author draws an explicit connection between human and stellar bodies, proclaiming, "Like us, stars are born, mature, reach old age and die."[27] Referring to Orion's stars as "mere toddlers," astronomer C. Robert Dell says, "When we look at Orion, we're seeing a star factory and what our solar system looked like in its infancy."[28] Imagining stellar matter as human origins, such discourses position Hubble images as if cosmic sonograms.[29]

But origins, Foucault reminds us, are never just there to be discovered. They, like human bodies and digital images, are actively made.[30] When Hubble images of the Orion and Eagle nebulae were released in 1995, news media celebrated them as if snapshots of human ancestors. Once again satellite television was being used in a practice of archaeology, but this time rather than excavating ancient ruins from the Mediterranean Sea, its gaze was poised to locate human origins in deep space. A *Newsweek* cover story entitled "Witness at the Creation" featured Hubble's view of the Eagle nebula accompanied by four smaller graphics illustrating the stages of stellar development. The article explains that the Hubble image reveals the stars' "extraneous gaseous globules," or "EGGs," and then goes on to claim that our universe "hatched" from EGGs like these.[31] These EGGs are "appropriately named," according to astronomer Robert Naeye, "because embryonic stars are forming within them."[32] In his book *Through the Eyes of Hubble: The Birth, Life, and Violent Death of Stars*, Naeye positions the Orion and Eagle nebulae as cosmic breeding grounds. In the chapter "A Star Is Born" the author evocatively describes the nebulae as fertile "stellar nurseries" attributing to cosmic matter humanlike reproductive capacities.

Emphasizing Hubble's unique ability to bring distant matter near, *Time* published an article entitled "Cosmic Close-Ups," featuring progressively tighter shots of Orion's "star factory" where stars are "mass-produced."[33] In a manner reminiscent of the Shoemaker-Levy 9 collision coverage, the writer treats Hubble images of star formation as analogous to Earth's own formation, speculating, "A similarly momentous and violent event (a shock wave of gas ejecting from a young star) probably accompanied the creation of our own sun and its solar system some five billion years ago."[34] Here Hubble's close-ups give expression to a turn of events so unintelligibly catastrophic that it can only be imagined in relation to the earth's creation. A similar *National Geographic* article, entitled "Orion: Where Stars Are Born," opens with a centerfold of "the heart of the Orion nebula" and declares Hubble's "unprecedented views of the nebula's center reveal the birth pangs of stars and perhaps the creation of planetary systems like our own."[35]

Hubble's images, so the logic goes, take us as close to witnessing the

Hubble images of the Orion nebula, which was described as "stellar breeding grounds."

The Hubble image of the tips of the Eagle nebula, which are described as a "star birth cloud."

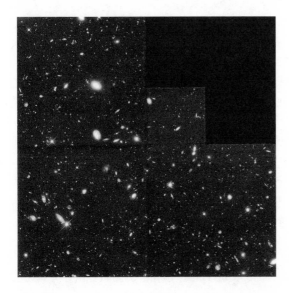

Astronomers used the
Hubble Deep Field image
to identify hordes of
galaxies and to delineate
the "edge of time."

earth's creation as we could ever hope to get. But if Hubble extends the gaze into deep space only to find it back in the cradle of the earth and human civilization, what does this suggest about televisual epistemologies? Perhaps the urge to transpose and move seamlessly across life and death, origins and catastrophes, beginnings and endings in the discourses surrounding Hubble images becomes a way of negotiating fundamental uncertainties that lurk within the televisual. These uncertainties are especially pronounced in sites of satellite television convergence that are defined by exceptional distance, time, and scale, and they often compel a discursive return to the flesh or the familiar.

Media coverage of the Hubble Deep Field image, which is said to have discovered a host of new galaxies, relies on tropes of human development to establish the image's significance. Referring to Hubble's images as "baby pictures," one news writer proclaims, "Hubble spied 14 primeval galaxies that seem to be the oldest normal galaxies ever seen. . . . Hubble is seeing galaxies as they were when the universe was a two billion-year-old stripling."[36] Another celebrates Hubble's ability to reveal galaxies "in the first blush of youth,"[37] and as Harvard astronomer Robert Kirshner speculates, "It looks as though there really was a kind of baby boom, a burst of star formation."[38] Astrophysicist Alan Dressler insists that the "Hubble Space Telescope, by telling us about how galaxies formed and evolved, is telling us something about how we got here."[39]

While it is typical for news media and astronomers to employ meta-
phoric language and analogies of human scale to explain scientific concepts
related to outer space, the excessive references to human reproduction and
development in the captioning and explanation of Hubble images suggests
there is something more ideological at stake. The farther Hubble looks into
deep space, the greater the tendency to anthropomorphize the images it
generates. As stellar matter is figured as a field of humanlike reproduction, Hubble images function not just as sonograms but as sites of cosmic
archaeology[40] and voyeurism, and our glee at discovering these "breeding
grounds" positions them as an interstellar primal scene—a peepshow into
cosmic copulation.[41] A primal scene fantasy, explains Constance Penley, is
"the name Freud gave to the fantasy of overhearing or observing parental
intercourse, of being on the scene, so to speak, of one's own conception."[42]
The Hubble Deep Field image, which peers to the "edge of time," has even
been described as a "keyhole" shot, evoking Freud's child illicitly snooping
on its parents' sex life.[43]

The suggestion that Hubble pinpoints human origins in deep space,
the belief that "we are star stuff," is symptomatic of anxieties generated
by this satellite-telescope-computer's capacity for fully automated and extremely remote viewing. In other words, the farther Hubble looks (or the
more televisual its gaze), the more uncertainty its images evoke, and the
more likely that they will be used to affirm the primacy of human life even
as the very definition of human life is itself increasingly up for grabs. Such
"astromedical" images, according to Sarah Kember, "compensate for the
declining status of the clinical gaze in the face of new imaging technologies (body scanners, for example) which can see both further and better
than the human eye, which unsettle the surface of representational realism, and which no longer centre and affirm the place of the humanist subject in the world which s/he surveys."[44] Since Hubble "can see both further
and better than the human eye," the satellite threatens to undermine the
authority of the scientist-viewer whose knowledge is derived through it.
Put another way, since the Hubble image implies the astronomer's obsoles-
cence, humanism must be constantly invoked within it as a way of affirm-
ing the human subject/scientific viewer sitting on Earth at this satellite-
computer-telescope's other end.

References to Hubble images as "star factories" also posit an interesting (if bizarre) comparison between Hubble and the hyperproductive stars

it makes visible. That Hubble's images are used to position stars as prod-

ucts of assembly lines in deep space also suggests the extent to which this
satellite-telescope-computer is implicated within the transition from an
industrial to a postindustrial economy, from an age of mechanical reproduction to an age of computer simulation. As a remote-controlled vision/
knowledge machine, Hubble is the perfect expression of an automated information economy: It constantly extracts light and heat from outer space
to produce and reproduce otherworldly matter as a continuous data stream
that can be stored, sorted, combined, doctored, and integrated within the
flows of global information economies. Any system of production (whether
stars in deep space or Hubble in Earth's orbit) that could be so extensive,
so prolific, and so essential can only be inculcated within an imaginary of
capitalist production. Perhaps ironically, then, the televisual epistemologies that form in relation to Hubble images tell us as much about how
tightly bound we are to earthly practices of carefully monitored human reproduction and capitalist production as they tell us about the contours of
the otherworld.

Cosmic Zooms

As Hubble images are absorbed in live media events and used to document
human origins, the televisual becomes a mode of distant observation that
gazes afar only to see what is near anew, a time machine that locates the
brink of history only to more clearly mark the present. There is, in other
words, a discursive elasticity embedded in this astronomical configuration
of satellite television. As Hubble images become cosmic events and cosmic sonograms, they place us more forcefully in our own historical moment and in our own bodies. Indeed, the very purpose of deep space viewing seems to be to affirm the centrality of the Western humanist subject
while disguising it as a process of searching for something other. Because
of this, catastrophic atmospherics can emerge but only in the living room,
and human-star genealogies can surface but only in bodies intelligible as
human. The discourses surrounding Hubble images work to negotiate the
ontological uncertainties and separation anxieties that are fundamental to
televisual ways of seeing and knowing and that are intensified by satellite
television convergence.

Televisual epistemologies are not confined to institutions of commercial or public television, or to astronomical science for that matter. Rather, as a set of knowledge practices they move between and beyond different sites of media and scientific culture and are derived through intermediale practices. The films *Cosmic Voyage* and *Contact* evoke Hubble's far-reaching vision with a digital effect known as the cosmic zoom, equipping the viewer with unprecedented visual mobility. Renowned as the "longest continuous computer-generated zooms ever created in the history of filmmaking,"[45] the cosmic zoom compresses fifteen billion years into four minutes and is designed to " 'fly' audiences to the outer reaches of space and then back into the smallest particles of matter."[46] Whereas *Our World* promised to zip the viewer on a tour around the planet, the cosmic zoom catapults us from a point on Earth to the edge of time and back again. As such it is symptomatic of the discursive elasticity that emerges in Hubble's cosmic events and sonograms and is consistent with a proclivity to appropriate the veneer of remoteness to confront something that is not just near but constitutive, interior, or fundamental.

The cosmic zoom was first developed for the NASA IMAX film *Cosmic Voyage* (1996), a science documentary exhibited at the National Air and Space Museum, distributed on video for space science educators, and aired on NASA-TV. The documentary was a collaboration by astrophysicists, cosmologists, and computer animation specialists, and like other NASA IMAX films such as *Destiny in Space*, *Blue Planet*, and *The Dream Is Alive*, *Cosmic Voyage* was produced for the eighty-foot-high screens and multichannel sound of the IMAX theater and aimed at "total sensory immersion."[47]

Adapted from Kees Boeke's book *Cosmic View* (1957), *Cosmic Voyage*, according to its creators, attempts to lead "the viewer on an exploration of the entire reach of the universe: from the smallest structure of the atom, out to the largest scales beyond superclusters of galaxies, and back in time to the early moments of the universe."[48] Its dizzying cosmic zoom sequences, since copied in science fiction films such as *Contact*, *Men in Black*, and *Sphere*, are accompanied by Leonard Nimoy's voiceover narration, which continuously reminds viewers of the dramatic changes in scale.

Nimoy initiates the first such sequence, asking, "As we look into the distant horizon, we may ask ourselves, what is our true place in the universe?" The cosmic zoom then flies the viewer on a visual journey from St. Mark's square in contemporary Venice (where Galileo invented the telescope in

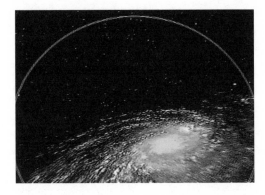

Cosmic Voyage structures a visual journey through deep space with the "cosmic zoom," which registers shifts in scale with concentric circles.

1609) through space to ancient spawning superclusters. The perspective zooms out continuously, from a single acrobat twirling a hula hoop, to a bird's-eye view of Venice, to a satellite image of the European continent, to an image of the entire earth. Concentric circles widening in powers of ten are superimposed on the image to establish a sense of scale. As the sequence continues, we move past the moon, and the tiny earth recedes in the distance as the entire solar system closes into view. We move back farther and see our sun "just as any other star." At "23 powers of 10" we glimpse the entire Milky Way galaxy. And then, suddenly, the zoom ends "15 billion light years from Venice," where "we approach the outer limits of the visible universe."

The cosmic zoom is constructed as the universe's largest dolly shot—an optical pullback that holds the earth in frame while the "camera" recedes until our entire galaxy becomes an infinitesimally tiny dot.[49] Thus while Hubble and all other telescopes necessarily peer outward, this shot imagines the fantastic view of looking back. We can only imagine and visually represent this perspective, however, because of Hubble's Deep Field image, which astronomers have coined to demarcate "the edge of time." As Nimoy reminds us, "What lies beyond this cosmic horizon, we cannot see and do not know." As the cosmic zoom shifts its perspective to a point where all vision and knowledge collapse, to a beginning "we cannot see and do not know," Hubble becomes the penultimate technology of archaeology, a device penetrating so deep that there is no time or space prior. The film's construction of the beginning as a dark abyss is only temporary, however. Since it insinuates such a profound displacement of Western science and humanism, both must be vigorously recuperated in a later sequence dramatizing the universe's evolution from the Big Bang to Hubble's launch.

The sequence opens in the palm of an astronomer's hand, where a tiny glass ball explodes graphically into flaming swirls of red, orange, and pink that transform into gigantic webs of gas, dust, and matter described as "the architecture of the cosmos." As this matter disperses, galaxies evolve into spirals and disks. Fuzzy balls of light emerge as stars, and we encounter our own Milky Way galaxy, where stars have formed and "died." A giant supernova bursts and "sends out the elements of life—the oxygen we breathe, the carbon in our muscles, the iron in our blood." We then recognize our sun, and see our own planet bombarded with cosmic dust and comets. Our gaze turns to the earth, where we watch an array of life forms emerge—green algae, wiggling crustaceans, swimming fish, crawling insects, soaring birds, roaming mammals, and eventually prehistoric men.

At the end of the sequence several dark-skinned hunter-gatherers appear, walking single file across a field with snowcapped peaks in the background. They arrive atop a cliff, where they settle for the night. With wooden tools one of the men builds a fire, and as he blows on a pile of tinder to generate a flame, the close-up of his face dissolves into a graphic match of the billowing orange explosion of Hubble's launch. This sequence is somewhat reminiscent of the Aboriginal video *Satellite Dreaming*, but instead of celebrating indigenous control over space technology, it constructs "primitive man" as a mere cog in the evolutionary wheel: The outcome of the invention of fire is the deployment of an enormous vision machine in outer space. As Nimoy explains, "Our cosmic voyage, from the Big Bang to the appearance of humans, took about fifteen billion years. From the beginning we were explorers, inventors, and technicians." He continues, "And in just a few thousand years, in just an instant of cosmic time, curiosity and technology would take us back toward the stars." We then see a shot of the orbiting Hubble Telescope through the window of the space shuttle *Discovery* as the earth looms in the background.

This sequence imagines the beginning not as a void but as a tale unfolding in the astronomer's palm, in primitive man's first fire, and in Hubble's launch. Interweaving Enlightenment science, Greek philosophy, and space technologies, the sequence forges an evolutionary time line in which past human inventions become the progressive preconditions for this "cosmic voyage." As the cosmic zoom etches this time line through deep space, it conjoins practices of astronomical observation and historiography along a linear trajectory where already seen and known phenomena can be rediscovered. Thus, while it may seem that the cosmic zoom would extend the televisual with its unparalleled visual mobility, *Cosmic Voyage* uses it in a way that reinforces linear structures and evolutionary logics, which harness televisual epistemologies to a trail of historical inevitability.

Despite *Cosmic Voyage*'s reduction of the televisual, the moving perspective of the cosmic zoom resonates with Walter Benjamin's discussion of the angel of history, which he identifies in a Paul Klee painting as "an angel looking as though he is about to move away from something he is fixedly contemplating."[50] The viewer in the cosmic zoom similarly moves away from the earth, mesmerized by its status as a miniscule dot in the cosmos. While we never see the viewer, we can only imagine the expression on his or her face while watching planet Earth vanish in its cosmic context. For Benjamin the angel bears witness to change, seeing history laid before his eyes, accumulated not in time but in space. The cosmic zoom, however, appropriates and transforms the space of history into a gliding linearity that interweaves cosmic evolution, human civilization, and scientific progress. For when the cosmic zoom's viewer looks back at the earth from the edge of time, he or she bears witness not to the unwieldy density of history but to a smooth line of scientific developments that purportedly made this perspective possible in the first place.

Whereas *Cosmic Voyage* uses the cosmic zoom to compress everything from the Big Bang to Hubble's launch into an expression of historical inevitability, the film *Contact* employs it in ways that complicate such structures of linearity. Adapted from Carl Sagan's 1985 book of the same title, *Contact* represents the possibility that the steadfast progression and rationality of the cosmic zoom might be disrupted or challenged by an astronomer, who functions more like an angel of history so dazzled by the density and complexity of historical accumulations that she is equipped to witness the unintelligible yet captivating matter of deep space.

The film opens with a cosmic zoom that begins with a view of Florida from the perspective of a passing satellite. As the sun rises over the horizon, the perspective pulls back past the earth and the moon, out of the solar system and eventually out of the Milky Way galaxy. Along the way we pass

Jodie Foster plays SETI
scientist Ellie Arroway in
Contact and uses the VLA
to detect alien signals.

landmarks visualized by Hubble, including stellar nurseries, globular clusters, and planetary disks. The visual trajectory from Earth to deep space is mixed with audio fragments from American and European film, television, and radio, slipping us further back into the twentieth century.[51] Finally, as we approach the threshold of the universe, the sound fades completely, and the perspective collapses in a blinding flash that dissolves to a sunlit window reflected in the iris of young Ellie Arroway's eye. Arroway, we learn, is a SETI astronomer in search of extraterrestrial intelligence. The cosmic zoom is used at the outset of the film to establish the vast domain of her search and to foreshadow her eventual voyage along that same path.

It is the sequence's random sampling of sounds, though, that is paramount; for it not only posits a nonhuman listener in outer space but also signals our inability to understand or control what extraterrestrial intelligence would pick up or decipher as importantly human. The sequence contrasts starkly with those in *Cosmic Voyage*, for rather than stretching linear models of Western civilization into deep space, it asks us to imagine what it would be like to hear our own past rearranged and read back to us. It invests an invisible, unknown, and displaced listener with the power to record and write the earth's history.

The relations between sound, history, and the feminine are layered in a later sequence again modeled on the cosmic zoom. This time Arroway herself goes along for the ride. During the launch Ellie is strapped into a capsule while scientists at the command center monitor her with audio, visual, and heat surveillance. When she plunges into space, the surveillance systems fail, and the scientists watch the transport device drop unevent-

During the astronomer's
journey through deep space,
she becomes a split subject,
unable to reconcile science
with pathos (Jodie Foster as
Ellie Arroway in *Contact*).

fully into the ocean. While the launch appears to onlooking scientists to have been an abject failure, we witness a remarkable voyage from Ellie's point of view. As her body is flung through several vibrant wormholes, digital effects literally split the female scientist's head in two. As a scientist Ellie describes what she sees—a four-star solar system, nebulae, transit systems, and other cosmic phenomena. But what Ellie *feels* is displaced into another persona—a visually split subject who erupts from the side of her face to blurt phrases like "I had no idea!" "They're alive!" and "Oh my God!" When the capsule finally lands, Ellie looks out a window and sees what she reports to be "some kind of celestial event." The image she sees is an actual Hubble image of the Eagle nebula. As a scientist Ellie attempts to describe what she sees, but her sensory immersion makes cool detachment impossible. She collapses in an emotional fit and proclaims in a shaken voice, "No words to describe it. Poetry! They should have sent a poet!" She repeats the phrase "It's so beautiful!" several times, overtaken with euphoria. The camera moves in to an extreme close-up of her eye and dissolves to an image of her body floating through space. She lands on a blissful hypercolorized beach, and the alien she encounters is her late father—here positioned as a "heavenly father." [52] Ellie is no longer a rational scientist, and her travel to outer space is figured as both a journey into her own past and a religious transcendence.

When Ellie returns to Earth, she is presented with visual evidence that the trip never physically took place and is ridiculed for losing her scientific objectivity. "I had an experience," she insists. But state and scientific authorities discredit her testimony at a public hearing, claiming her ac-

count was memory based, emotional, and subjective. The command center observers insist that she never left the planet, although we later learn
that a black box recorder inside the capsule registered eighteen hours of
static. The disjuncture between her story and official skepticism leaves the
female scientist in a profound state of confusion and contradiction. For
what can she resolve, if not an inquiry that takes place within her own
subjectivity?[53]

Contact uses the cosmic zoom to explore the limits of outer space and
human knowledge, but these limits are ultimately articulated within the
feminine as an inability to see "objectively" and with appropriate distance—
an inability to reconcile the physicality of the body with the dominant
scientific order. In Contact feminine physicality introduces the possibility
that the linear flow and rational unfolding of the cosmic zoom might
be disrupted. Although the film's narrative imposes fierce sanctions on
Ellie's embodied view, her space flight nonetheless serves as an important
intervention into the imperializing prerogatives of Western science. Like
Hubble itself, Ellie moved beyond the murkiness of the earth's atmosphere
to glimpse the cosmic horizon, but what she "discovered" was a refusal to
accept her views—views that were fundamentally about the inability of language to make sense of something other.

We might interpret the recorded static as the unknown historian's will-
ful erasure of a time beyond Earth, as a way of signaling the lack of human
language, imagination, and representations necessary to account for the
difference of outer space. This rings especially true if we recognize that uses
of a vision machine as powerful as the Hubble Space Telescope have not
necessarily spawned new ways of seeing and knowing in our culture but
have resulted in a compulsion to pull what it senses through the prism
of what we already claim to know. While it would be impossible to fully
escape the discourses and languages through which we come to know, it
seems equally crucial that we find ways to imagine the light, heat, and matter that Hubble senses as nonhuman, digital forms of an almost incomprehensible scale.[54] This, of course, is not a technical problem but rather
one of the Western imaginary and its stubborn ties to scientific rationalism
and Christianity addressed in Contact's narrative. So long as there persists
an unwillingness to imagine and accept difference in astronomical fields,
Hubble will continue to show us only what we can bear to look at, whether

the catastrophic disaster of another planet, the humanlike productivities of stellar matter, or Earth's central position in the cosmos.[55]

Cosmic Signals

The incommensurable relations between astronomical observation and *difference* that flare in cosmic events, sonograms, and zooms also surface in the signal exchanges that organize the narrative of *The Arrival*. Like *Contact*, this film thematizes televisual epistemologies, but instead of using the astronomer's views to challenge scientific rationalism, *The Arrival* features SETI scientists who rig various astronomical technologies to deter alien invasion and patrol borders on Earth.[56] Their work commences at the VLA (Very Large Array) in Socorro, New Mexico, a massive radio telescope observatory comprising twenty-seven dish antennas, each with a diameter of twenty-five meters, configured in a *Y* formation to collect radio signals emitted by astronomical bodies.[57]

In an early scene renegade SETI astronomer Zane Zaminski and his nerdy partner, Calvin, work yet another late-night shift, scanning the heavens at the VLA. They are jolted into action by a startling spike on their computer monitor, and they record a forty-two-second "non-random, non-earth-based" signal in the 107 megahertz range (FM radio band), which they deem a sure sign of alien intelligence. Zane presents a recording of the noise, which he thinks might be a "distress signal" or "alien encyclopedia," to the director of NASA's Jet Propulsion Lab, who immediately fires him, explaining, "The search for ET's in this political environment is a tough sell on Capitol Hill." (It turns out that this NASA boss is really part of the alien conspiracy.) Resourceful and persistent, Zane lands a job working for the local direct satellite broadcasting company and roams door-to-door to perform a "courtesy service" for his white suburban customers. Little do they know that he has linked the neighborhood's direct satellite broadcasting dishes by fiber optics to create his own very large array. Using this makeshift contraption Zane intercepts the same strange signal, but this time it is in the unlikely guise of a pop song emanating from a radio station in central Mexico. As we soon learn, this radio broadcast is merely a cover for a most malignant form of "border blaster," for a hostile alien species is busily colonizing the earth. The signals Zane detects are encrypted communications with the home planet, and their new command post is a defunct power plant in rural Mexico.

When Zane travels to Mexico to investigate the source of the signal, he encounters Elana Green, an atmospheric scientist studying global warming. After analyzing years' worth of remote sensing data, Elana discovers that the emission rate of greenhouse gases in central Mexico, Ecuador, and Brazil has skyrocketed 700 percent during the last five years, and, as a result, the earth's temperature is projected to rise twelve degrees centigrade in the next decade. Together, Zane and Elana use their intercepted signals and satellite images to unravel an alien conspiracy (which has also infiltrated NASA) to accelerate global warming and terraform the earth's atmosphere so that it is fit for alien life.

Whereas in *Contact* astronomical observation is refigured as an emotive scientific practice, in *The Arrival* it becomes one of global security. Zane and Elana commandeer a medley of direct satellite broadcasting, remote sensing, and radio astronomy technologies to attribute the problem of global warming to a Hispanic-looking alien civilization that has retrofitted defunct Central and South American power plants to pump out vast quantities of greenhouse gases. A sequence of colorful animated satellite images displays temperature data across the Northern and Southern hemispheres, contrasting the hot, polluting world of the developing South with the relatively cool, clean world of the postindustrial North.[58] This seemingly remarkable assertion—that the developing world is solely responsible for global warming—is, of course, matched only by the fact that within months of this film's release, at an environmental conference in Kyoto, Japan, representatives of Western industrial nations made essentially the same claim.[59]

In addition to its rash spin on global warming, *The Arrival* allegorizes white American anxieties about Mexican immigration in the United States. Aliens are disguised in Hispanic-looking brown skins, their intelligence is camouflaged as Mexican popular music, and an array of satellite dishes in the white suburbs of the Southwest are mobilized to detect their presence south of the border. The film's representation of alien invasion resonates with reactionary efforts (in California and elsewhere) to reduce immigration quotas, enact English-only education laws, and militarize the

borders.[60] In short, the film collapses an imminent extraterrestrial invasion
onto white conservatives' anxieties over Mexican border crossing.[61]

In *The Arrival* radio astronomy, direct satellite broadcasting, and remote
sensing become border patrol technologies. Radio astronomy is used to detect alien intelligence in deep space; direct satellite broadcasting is used
to detect aliens in Mexico; and remote sensing is used to identify developing nations as the earth's pollutant "hot spots." Almost echoing the discourses of *Our World* decades earlier, satellite television (whether remote
sensing, radio astronomy, or satellite broadcasting), the film implies, is a
crucial means of maintaining Western hegemony over a "dangerous" third
world.[62] As the film conflates anxieties about immigration, global warming, human evolution, and alien life, it represents astronomical observation as patrolling boundaries between Earth and outer space, first and third
worlds, human and alien. Whereas *Contact* represented the possibility that
astronomical observation might detect a listener in outer space that could
intercept our signals and deflect them back to us, *The Arrival* forecloses
that possibility. Instead, it suggests that the gaze into space is ultimately a
ruse since "aliens" have already "arrived" on the planet, and the most urgent
concern is to redirect the world's most powerful vision machines to police
the borders of humanity on Earth. The problem, of course, is that these
borders are strictly determined by the logics of Western Eurocentrism.

The Arrival is a highly symptomatic text in that it recombines and
concretizes discourses alluded to throughout this chapter. Moreover, it
clarifies the way astronomical observation and technologies are embedded
within a broader system of cultural politics. Just as Hubble images of
Shoemaker-Levy's crash were used to emphasize the fragility of Earth,
The Arrival foretells of our planet's imminent colonization. Just as Hubble
images of stars were cast as human origins in deep space, Zane and Elana
used devices to define and defend "humanity" on Earth. And just as Hubble
images and the cosmic zoom have been used to celebrate linear models
of Western technological progress, the film suggests that with enough in-
genuity any combination of technologies can be repurposed to protect the
industrial society that yielded them. The film's themes, in other words,
resonate with the preoccupations manifest in the discourses surrounding
Hubble images.

Both cinematic representations and news media are part of the social construction of the Hubble Space Telescope. They form an intermediale network of texts that shape our understanding of Hubble and its relation to televisual epistemologies. Only by considering the different media contexts in which Hubble images are integrated or invoked can we understand how the device functions as satellite television technology and how it generates cultures, histories, and sciences in orbit. Just as our understanding of Hubble is shaped by news media and cinema, "satellite television" is constituted in part by the panoramas mapped throughout this chapter. These panoramas are the critical spaces symptomatic of satellite, television, and computer convergences. They are also the muddled terrains where the abstraction and unintelligibility of space matter are constantly negotiated with globalizing claims to knowledge and the tendencies of Western Eurocentrism.

Conclusion

The visual mobility afforded to the Western viewer by technologies of astronomical observation can be understood as one of the cultural spin-offs of the military-industrial-information complex. This moving perspective is symptomatic of immense military, corporate, and scientific capital invested in television, satellite, and computer technologies over the past several decades. Perhaps more than anything, the cosmic events, sonograms, zooms, and signals discussed in this chapter reveal that Hubble has been used not only as a technology of astronomical observation but also as one of remote control. In television studies this term has multiple meanings.[63] Television, according to the editors of *Remote Control*, "controls us at a distance. It emanates from some place far away, yet it makes its presence constantly felt in our everyday lives."[64] When articulated through Hubble's far-reaching gaze, the strategy of remote control treats outer space as an astonishing realm of new frontiers, new discoveries, and new knowledge only to tether the events and matter Hubble detects to Western civilization and scientific progress, only to rewrite the bounds of the "historic" in ways that reinforce linear developmental narratives of human evolution, only to patrol and maintain imaginaries of Western Eurocentrism. Remote control, in other words, involves a compulsion to bind whatever Hubble's gaze

"discovers" out there to Western humanism, and it emerges most forcefully
in the public presentation of these scientific images, whether in the positioning of Hubble images as "live coverage" of another planet's catastrophe,
as anthropomorphic renderings of stellar matter, or as otherworldly views
to be decoded only within scientific rational frameworks.

Thus television is not just an ideological system in a cultural Marxist
sense. It is also an epistemological system in a Foucauldian sense. By examining its convergence with satellite technologies, I have explored how
it is used to produce, assemble, organize, and order a range of knowledge
practices. Remote control is a useful metaphor because it evokes meanings of the televisual that have been derived in relation to scientific and
military paradigms as opposed to distinct from them. As the discussions
throughout this book suggest, the medium's convergence with satellites
and computers positions it not just as a form of commercial entertainment
but as a technology of surveillance, archaeology, and astronomy as well. In
other words, satellite television is driven as much by military or scientific
prerogatives as by commercial gain or public education.

If television functions as "a window on the world," then the power to
remotely control or regulate what appears in view is, to a certain extent,
the power to delimit and determine knowledge of the world and to selectively shape the contours and meanings of it. Just as television viewers use
the remote control to hone in on particular views from the living room,
so, too, do astronomers use Hubble to target and produce images of particular sites in deep space. Although one practice is motivated by a desire
for entertainment and the other is guided by scientific investigation, both
involve uses of technologies to see, know, and control remotely. And both
involve knowledge practices that require the viewer to recognize vast temporal and spatial distances, whether the ancient past or deep space, while
enabling him or her to simultaneously feel at home.

Shoemaker-Levy's collision with Jupiter became a "live" media event
not unlike the early satellite spectaculars. But as Hubble connected viewers in near–real time to a catastrophic event on another planet, it also enabled a form of distant monitoring that resembled the passionate detachment of U.S. satellite coverage of the war in Bosnia and the North/South
divisions perpetuated by *Our World*. In other words, it offered a comfortable platform for viewing the distant devastation or trouble of others. This

comfortable view is not innocent. It marks the way the industrial West has devised and used various combinations of "satellite television" to reassert its global hegemony in the context of decolonization, outer space exploration, post–cold war reorganization, and global digitization. The West's neoimperializing and neocolonial forms of power rely on uses of televisual technologies to draw lines around the planet and to patrol and defend the geopolitical and epistemological boundaries of the West. *National Geographic* audaciously exemplifies this neoimperialism in its celebration of Hubble: "Thanks to Hubble we can begin to register the notion that while our earth is our local address, we have an entire universe that we can call home."[65] Despite such statements, particular constellations of "satellite television" can bring geopolitical and epistemological categories into relief and complicate them, ranging from the "global now" to the "satellite dreaming," from "ancient trysting grounds" to the "edge of time."

On November 2, 1995, CNN evening news broadcast a segment that included Hubble images of newborn stars emerging in the Eagle nebula. After the segment aired, viewers flooded CNN's phone lines to report having seen a religious image in one of Hubble's pictures. The next day, CNN *Today* took calls from viewers who had allegedly witnessed the face of Jesus in Hubble images of the Eagle nebula. A caller from Texas proclaimed, "When I walked by the TV and looked at that I thought, 'My God, they say this is the birthplace of stars and things, and that appears to be Jesus Christ.'"[66] The alleged visibility of such otherworldly creatures in and around Hubble images is part of an important social struggle over this satellite's televisual and scientific systems of knowing.[67]

Hubble images are more open to such popular skepticism than most kinds of visual evidence for several reasons. First, Hubble images are generated from the perspective of a remote, orbiting satellite, from a position that no human eye can occupy; thus there is virtually no way to verify or authenticate what lies in its field of vision.[68] Second, Hubble images claim to represent massive matter never known to exist before through almost incomprehensible distances. They reproduce faraway matter of enormous dimensions and in so doing decenter humanist understandings of distance and scale. Finally, Hubble's images are digital and can be reproduced and undetectably manipulated; thus their relationship to the materiality of outer space is constantly called into question. This combination

of characteristics has the effect of arousing impassioned efforts to ground
these images within world history and in relation to the human body, hence
their incorporation within such discourses as the live media event, the
sonogram, and the zoom.

Still, Hubble's satellite panoramas cause us to rethink the determinacy
of visibilities because they are so disembodied, remote, and abstract—so
very indeterminate. Once again, these satellite images share more with abstract expressionist painting than they do with the culture of photographic
realism. As a result, they may be more malleable and open to cultural use
and interpretation. Despite this potential, however, satellite images have
historically been used by state and scientific institutions as an almost tran-
scendent form of knowledge and power, as the most objective view, the
most official kind of visible evidence. It is this disjuncture between the
firmness or clarity of scientific objectivity and the blurry contours of satellite images that interests me. For this contradiction, I believe, generates
an important means of contesting the ways that satellites are used to see
and the kinds of knowledges they are used to produce. To signify anything
other than their claim to omniscience, satellite images must be put into
discourse. It is the necessity of interpretation that's so important here, for
unlike other forms of visual culture the satellite image is not self-evident,
despite its associations with omniscience and objectivity. In an age in which
everything appears to be visible (from the "edge of time" in deep space to
the human DNA that makes up our bodies), satellites require us to make a
provocative turn to unknowing, which keeps practices of reviewing, interpretation, and media literacy alive.

In all of its forms satellite television is a technology of cosmology—
an active producer of narratives of human, world, and cosmic history. Because of its imperializing potential it is crucial that we look to alternative
(non-Western) knowledges and views formed via satellite. In controlling
satellite technology Aboriginal Australians, for example, do not directly
control the meanings generated by the astronomical observatory in Parke,
Australia, featured in *Our World*, nor do they direct the gaze of the Hubble
telescope. They have, however, used Imparja's satellite access to circulate
countermemories of the creation of the universe, offering an alternative
cosmology that contests Western knowledges produced via satellite.[69] Indeed, the dreamtime may posit an altogether different account of the uni-

verse's formation. By struggling for control over satellite technology, then, Aboriginals gain the power to circulate their own narratives about the earth's formation and their territories in the outback. Such struggles are crucial, because they represent uses of satellite television to contest the official truths by which Western military and scientific interests work to maintain their hegemony.

Although Hubble images document outer space phenomena, they can only be interpreted at all *in relation to* what we more intimately know—that is, histories, societies, and cultures on the earth. When satellites extend the televisual gaze around, beyond, and beneath the earth, asserting its presence globally and cosmically, we must ask, Who is present? Who is seeing? Who is knowing? Such questions will help us to understand and explore how television operates, not just as a complex institution of broadcasting, program production, and reception but also as a broader epistemologic system that draws on and in turn reshapes a wide range of scientific, military, and cultural practices. In astronomy the "event horizon" refers to the boundary region of a black hole—the point beyond which nothing, not even light or knowledge, can escape. With its gaze directed toward deep space, Hubble helps to establish what we might call a cultural "event horizon," exposing the brink of the televisual, the edge of epistemology, and the limit of remote control.

Conclusion

I began this book with a description of my first encounter with a satellite, and I have attempted to amplify this moment with a multitude of satellite crossings, charting a medley of satellite uses, generating a genealogy that skips across centuries, disciplines, and continents. Although the transformations facilitated by satellites have been gradual and indeterminate, many were foreshadowed in the first experimental satellite broadcast, *Our World*, which stands out as a prophetic text and marks a certain point of departure. *Our World*'s visit to an Australian radio astronomy station, its portrayals of folkloric dancers and musical performers, its integration of a remote-sensing image from the ATS-1 satellite, and its emphasis on Western modernity resonate with other satellite uses explored throughout this book. *Our World*'s treatment of these wide-ranging phenomena was symptomatic, in other words, of the ways the technology would be used in decades to follow. This brings me back to the initial impulse of this project, which was to determine how satellite uses have reconfigured our encounters with and understandings of television. So, what is television in the age of the satellite? As a way of answering this question I want to highlight four themes running throughout the book.

First, *satellite television is a site of technological convergence*, which means that it is not one but several technologies that have come together in different ways at different times. In this study I have examined *discursive sites* as opposed to technological hardware in an effort to push the study of convergence beyond questions of function and to engage issues of meaning and power. In addition I have attempted to interweave questions of convergence with those of genealogy as a way of stressing the disparate kinds of uses and practices through which satellite television has been constituted.

Through these discursive sites I have tried to delineate the technological practices of live international broadcasting, direct satellite broadcasting, remote sensing, and astronomical observation. Since the satellite is so often imagined only as a distribution device for broadcasting or within the purview of military or scientific institutions, it is rarely considered as part of television's content and form. But satellite television convergence resulted both in the technological practices mentioned above and in the emergence of new audiovisual formats, whether the live spectacular, the hybrid flows of Imparja TV, the orbital coverage of war in Bosnia, the satellite and digital interfaces of archaeological excavation in Alexandria, or the deep space panoramas of the Hubble Space Telescope. All of these formats can be considered part of television's changing content and form, which shift not only with sociohistorical conditions but with processes of technological convergence as well.

A convergent model of television involves doing away with mind-numbing metaphors such as the boob tube or the idiot box, which are intended to simplify television and its watchers, and imaginatively replacing them with more lively, eclectic, and chaotic ones such as the sculptures and installations by Nam June Paik. These works are convergent in a most physical way, configuring satellites, computers, and television as entangled parts, at the same time cleaving open the boundaries between television and art.[1]

It is also important to note that a convergent approach to television technology is not unique to the digital era. It was embedded in the writings of earlier thinkers such as Raymond Williams, Marshall McLuhan, and others who looked both ways, so to speak, understanding television as both derivative of previous technologies and as one of future transformations and potential. As Langdon Winner reminds us, "New technology . . . typically emerges not from flashes of disembodied inspiration, but from existing technology, by a process of gradual change to, and combinations of, that existing technology."[2]

In the context of digitization the issue of convergence has beamed brightly on the radar screens of media scholars invested in maintaining distinctions between various mediums or formats, since the computer represents the potential to seamlessly blend them. While it is important to focus on media specificities, it is equally important that we not become so preoccupied with making such distinctions that we end up inadvertently writing ourselves out of crucial discussions and debates about the development and uses of new media and new televisions. A convergent approach

to television involves keeping the meanings of the technology dynamic and
malleable, open to being mobilized and used in different directions, across
languages and disciplines, and in unpredictable ways. It also involves rewriting our critical terms and keeping them useful as television combines
with and is altered by new technologies. There are many televisual precedents for current practices that are too often understood only as "new" or
"digital" media. We cannot afford to be purists about defining what television is, which is a challenge given the newness of our field and the need
to specify, historicize, and legitimate it as a viable area of study. We do ourselves a disservice, however, if we draw narrow lines around a medium that
is so broad.

Second, *satellite television is a technology of knowledge*. It is an epistemological system structured in part through the technologization of sound
and vision. To say that satellite television is a technology of knowledge
is to recognize the array of knowledge practices within which it is implicated and to foreground its constitutive relation to the formation of ways of
seeing and knowing the world. As satellite television practices have been
operationalized, the very definition of "the global" has increasingly become
contingent on them. I do not mean to suggest that satellite television fully
determines worldviews but to emphasize, instead, how its practices are
structured to make the world intelligible in certain ways. I argue that the
televisual should not be reduced to commercial or public service broadcasting. Rather, the televisual involves the production and circulation of knowledges through the different discursive modalities of commercial entertainment, public education, military monitoring, and scientific observation.
The televisual, in other words, take shapes as the articulation and combination of these modalities, which are rubrics through which knowledge and
power are structured, produced, and distributed.

For instance, by hopscotching around the globe to showcase cultural
performances and by focusing on the population explosion, *Our World*
structured knowledge primarily through the modalities of public education and scientific observation. It constructed the planet as a domain that
could enlighten and uplift, as well as one of considerable risk that had to be
monitored and managed. Hubble's images of stellar formation emerged for
purposes of scientific observation and public education but became com-

mercial entertainment when Hubble images appeared in TV news and science fiction drama. Satellite images of war in Bosnia first served as a form of military intelligence, then as commercial television news, and finally, after being subjected to scrutiny and discussion, a site of public education and witnessing. Conceptualizing the televisual in such a way requires remaining aware of the variable and shifting knowledge practices articulated through television content.

By treating television as an epistemological system we also open our object of study to a broader range of interdisciplinary dialogues, technological arrangements, and political interventions, which may spur critiques of the medium across geographic, technological, and disciplinary boundaries. Perhaps most important, we might explore how the televisual functions as an imaginary itself (a function due in part to the historical process of its naturalization as "global presence"), permeating and reconfiguring times, spaces, objects, and bodies in different ways at different historical moments. I don't mean to offer a totalizing description of television or to imply that television is essentially a form of omnipresence or omniscience. Instead, I am suggesting that precisely because of television's historical uses to assert and maintain the dominance of the industrial West over the air, across the land, through orbit, and on the screen, we can assume that it *might be* found in the most unlikely places. If this is the case, then we need a critical model that takes into account the way television takes shape as a system of knowledge and power that can be uplinked and downlinked, archived and circulated, displayed and concealed. These practices occur not only within institutions and across screens but also through territories and the various parts of the apparatus. Television is where critics and historians want to find it—and if we look for it in unexpected places, then perhaps we will gain some understanding and control over the way it has knowingly and unknowingly been used to produce and circulate our own views and knowledges of our world.

Third, *satellite television can and should be used by subordinated social formations, artists, and activists.* Some may find it ludicrous to think about the appropriation of satellites by subordinate social formations, but it is just as important to imagine alternative uses of satellite television as it is to recognize the ways in which they have been used to assert and renew Western hegemony. Critics worry that the eight thousand satellites now in orbit will only further elaborate what Anthony Giddens has called "the totalitarian tendency" of the state in late capitalism, especially since most are owned by the United States and Russia. Since the 1970s, however, media artists have been challenging the military, scientific, and corporate authority over these space-bound machines.[3] For over a decade organizations such as Imparja TV and Deep Dish TV have used satellites to circulate indigenous or counter-cultural programming, and a handful of artists have used them to create artworks as well.[4] In 1977, long before Web-based art, American performance artists Kit Galloway and Sherrie Rabinowitz developed *The Satellite Arts Project* and used satellites to explore the possibility of "telecollaborative art" and "virtual performance spaces." Sponsored by NASA, the NEA, and the Corporation for Public Broadcasting, the artists used live satellite transmission to explore what they described as a "new way of being in the world."[5] In one sequence, for instance, two bodies dancing on different continents appear to be holding hands within the same frame, and the words "3000 miles apart" are keyed between their bodies. Such images, they explain, "demonstrated that several performing artists, all of whom would be separated by oceans and geography, could appear and perform together in the same live image."[6] What I want to emphasize here is the significance of a media-performance art practice that imagines itself not only as a struggle over representation but as synonymous with the satellite apparatus—as part of the satellite's potential for the global circulation of signals. The work of these early orbital performers is also important because it deterritorializes artistic practice, offering a decentralized and orbital model of cultural production that is manifest as dispersed performing bodies, stages, satellite transponders, and television monitors. Galloway and Rabinowitz describe their project as an attempt to use satellites and video to generate "a performance space with no geographic boundaries."

Media activist Brian Springer has also used satellites in innovative ways. In 1992 he set up several satellite dishes at his home and began downlinking and capturing the raw satellite feeds of commercial TV news networks in the United States. On these "backhauls" one could see TV personalities and political leaders during the time between their live appearances on different news networks as they are being made up, cajoling, primping, or whispering. As Springer explains, "To the networks, these feed out-takes are trash, and to most home dish-owners boring. To me the feeds are a

Makrolab is an "outpost for tactical media work" that looks like an Earth-bound satellite. Courtesy of Marko Peljhan.

window onto the construction and performance of character." Springer recorded more than five hundred hours of these feeds and compiled some of this footage into a 1996 video he called *Spin*, which details how political leaders use satellites to "tour the world electronically," efficiently showcasing their political presence. *Spin* exploits techniques of downlinking, archiving, and recontextualizing to expose the time and space between the generation of a raw satellite feed and the packaging of it as part of commercial television flow. By focusing attention to raw satellite backhauls, Springer also produces knowledge about what satellites can and do deliver: series of carefully contrived performances in which everything from political speech to makeup is regulated or "spun."

Finally, Slovenian artist Marko Peljhan is the creator of a project called Makrolab. Described as "an outpost for tactical media work," Makrolab is a mobile unit that looks like an earth-based International Space Station. It is moved to a different part of the world every one to two years and inhabited for three months by scientists, artists, and researchers working on collaborative interdisciplinary projects. The project began in Germany in 1997 and has since landed in Australia (1999), Scotland (2002), and Italy (2003), and it is scheduled to end up in Antarctica in 2007. Always stationed in remote locations, Makrolab relies heavily on satellites and telecommuni-

cation networks. During the 2002 installation in the Scottish highlands,
for instance, resident Makronauts used high-speed satellite Internet access, satellite tracking software, satellite television receivers, and global-
positioning satellites. Peljhan describes Makrolab as both a sociological
experiment and an opportunity to "live in the band" (bandwidth, that is).
As he explains, "if you live in the spectrum, then you have the direct experiences of what is happening in the band. You start recognizing details,
also very intrinsic ones, hidden ones, operational plans and modes, you
start understanding how somebody or some larger system is using the
spectrum."[7]

In 1997 Peljhan and Springer worked at Makrolab and conducted satellite signal reception and interception exercises, which later became a
more general "Electronic Media Monitoring" project. They used these satellite signals to produce a CD entitled *Signal Territory* (2002). Much of
the CD includes electronic musical compositions that sometimes sample
"found" signals, but there is also a highly significant forty-four-minute
track called "Sector V," which makes five raw satellite intercepts available.
It includes conversations, for instance, of U.S. military personnel devising
new weapons systems and international officials planning interventions
in Sierra Leone's civil war. Peljhan hopes that the CD will help listeners to
understand the kind of content that passes across the surfaces of satellite
transponders and dishes and provoke them to begin asking "who speaks
through, owns, controls, and dominates the invisible yet ubiquitous space
of the signal territory?"

Peljhan is one of the few artists who has adopted an orbital work model.
All of his projects involve signal reception and redistribution, operations
across geographic domains and languages, international collaboration, and
limited duration. In Peljhan's work the satellite is integrated in a material
and structural way. He has, in effect, used the Makrolab as a giant receiving
device and remote sensor, inviting international participants to live and
work there; receive, store, and rearrange signals; engage with and sense
the surrounding physical and social environs; and leave a trace in the form
of a project. In this way Makrolab is a radical interpretation and artistic
appropriation of the *Our World* concept.

The Satellite Arts Project, Spin, and Makrolab encourage viewers/citizens/

consumers to become more knowledgeable about the technical properties and current uses of satellites and to reimagine, reinvent, and retool
them in ways that are geared toward demilitarization and decorporatiza-
tion. We need more such works that imagine and suggest ways of struggling over the meanings and uses of various forms of satellite television.
While the industrial West may dominate the satellite television scene, it is
vital that we refuse technologically determinist arguments even in relation
to technologies that seem so patently beyond our control. Works like these
may provide just the kind of political spark needed to ignite new orbits of
humanist intervention, media activism, and artistic appropriation around
the satellite.

Finally, *satellite television is a part of an ongoing dialectic of distance and
proximity.* By this I am referring to television's capacity to produce a structure of feeling that enables an experience of simultaneous connection and
separation. Television's convergence with the satellite has, I believe, intensified and reconfigured this dialectic in multiple ways. In the 1960s
satellite spectaculars offered Western viewers an experience of global presence, instructing them they could be anywhere and everywhere in the
world simply by watching the TV monitor. By the 1990s, however, satellite
television had a more sinister configuration of global monitoring, encouraging Western viewers not only to experience the world from a distance
but to imagine themselves as safely guarded from atrocities throughout it,
whether in Iraq, Bosnia, Rwanda, Kosovo, or Afghanistan. This dialectic of
distance and proximity can, after all, take shape as an extension of colonial
practices of travel and global crossings. In some instances remote-sensing
satellites have been used to further mitigate anxieties surrounding "close
encounters" and "contact moments" by fostering the formation of "secure
perspectives," views of troubled peoples and territories seen from a safe
distance.

As I have suggested, however, the uses of remote sensing are inconsis-
tent and paradoxical. In the case of the Alexandria excavation archaeologists mobilized televisual technologies to gratify a desire for proximity and
tactility with Cleopatra. Here not only is the dialectic eroticized, it also assumes a temporal dimension as the spatial metaphor of proximity and distance is transposed onto present and past. In this context, then, a Western
patriarchal version of the dialectic emerges as satellite television practices

are combined to fulfill heterosexual male desire, advance archaeological
science, and write ancient history. If we extend this discussion to Imparja
TV in Australia, however, the dialectic is reconfigured yet again. Aboriginals
living in remote areas have used satellite television to sustain community
autonomy while remaining globally present and locally unified through TV
signals that traverse Aboriginal lands. In this context the meanings of distance and proximity are interwoven with postcolonial tactics of cultural
survival.

In his book *Speaking into the Air* John Durham Peters suggests that all
communication at a distance is "an expression of desire for the presence
of the absent other," and he frames this dialectic as a "quest for authentic
connection."[8] While television, no doubt, often functions as a medium of
connection and even intimacy, I would argue that it should not inevitably be
conceived as such. The dialectic of distance and proximity, in other words,
should be considered and analyzed in relation to broader power relations
(whether postcolonial or gender conditions) and the various discursive modalities of the televisual. Resisting technological essentialism enables us
to better understand how the meanings of this dialectic may shift with the
unpredictable and contradictory uses and effects of satellite television technology. The further the televisual gaze looks into deep space or the deep
past to find something other, the more forcefully it may be bound to Western civilization and human reproduction. The more unity that is celebrated
in the live global television relay, the more divisiveness it may implicitly
create. The more abstract and distant the satellite image, the more tactile
and sensuous the forms of engagement it may necessitate. Clearly, the uses
of satellite television have rendered the meanings of connection and separation more political, multivalent, and complex.

Global Positioning

Satellite television's dialectic of distance and proximity became acute and
personal for me as I was writing this manuscript on September 11, 2001,
the day of the attacks on the World Trade Center and the Pentagon. In a
state of concern and distraction I worried about my family and friends in
New York and felt a sense of proximity only by virtue of the television signals ricocheting live via satellite from one coast to another. I morbidly won-

dered if there were satellite images of the crash sites and later discovered there were many. And I felt a little strange given that the same symbolic targets that were attacked—U.S. global capitalism and militarism—were also implicitly critiqued throughout this book. But this moment became relevant to the closing of this book for several other reasons as well. For it was primarily through live satellite feeds, remote-sensing images, and satellite-relayed phone calls and email messages that citizens, like myself, tried to make sense of the rarity of these historic events and position ourselves within them.

As I watched the towers collapse on TV that morning, I realized that the multiple discursive modalities of the televisual were being mobilized in full force. Suddenly commercial entertainment, public education, scientific observation, and military monitoring collided in this coverage that lasted not just for days but months. Television news broadcasts beamed live-via-satellite views round the clock from New York and Washington, D.C., informing the public about the attacks and their aftermath. In the days that followed, *Ikonos* satellite images of the World Trade Center and the Pentagon appeared on television, in newspapers, and online and served as scientific maps of destruction used in damage assessments and artworks alike.[9] The 9/11 attacks were notorious for interrupting the flows of commercial entertainment, whether late-night TV talk shows, Disneyland rides, or Hollywood movie releases, but they actually revitalized satellite television news networks as viewers clustered round to watch history unfold in the present. The demand for post-9/11 news pumped life back into CNN, sent the ratings of the Fox News Channel skyrocketing, and put Al-Jazeera on the global media map. The event created opportunities for lesser-known cable channels as well. The Oxygen network, for instance, orchestrated a series of rare live satellite broadcasts in which women from across the United States and around the world commented on 9/11 and evaluated U.S. plans for military retaliation. Oxygen's two-hour special "Women of the World Respond" linked women in Sarajevo, Johannesburg, Paris, London, Pakistan, Moscow, and New York live via satellite, presenting a range of women's perspectives, many of which ardently condemned U.S. militarism and encouraged peaceful alternatives.

When the United States finally did retaliate against Osama bin Laden, the Taliban, and Al Qaeda in October 2001, the State Department comman-

Ikonos satellite image
of the World Trade
Center on September
12, 2001. Courtesy of
Space Imaging, Inc.

deered all satellite images of Afghanistan, preventing refugee workers,
medical and emergency crews, and journalists from accessing and using
them in relief efforts or war reporting. By late November, however, the State
Department, in a manner reminiscent of Bosnia, declassified select before and after satellite images confirming that many Taliban and Al Qaeda
targets had been destroyed once and for all. Fox News and CNN rushed
to cover military briefings and press conferences in which Bush administration officials proudly displayed satellite images and cockpit videos of
missiles pummeling targets. These forums appropriated the televisual to
advance a state-led paradigm of safe-distance annihilation and revenge.

Further reinforcing television's militarization, CNN used its satellite
time to air shows like Wolf Blitzer's "Military Options," which gave little, if
any, time to options other than military ones and privileged instead a U.S.
nationalism that altogether negated possible peaceful solutions.[10] In an act
of network vigilantism, Fox News deployed renegade journalist Geraldo
Rivera to the front lines of the Afghani resistance, showing him live via
satellite crawling around the desert in military garb, befriending Afghani
fighters, and fending off bullets in the crossfire. Rivera became a vengeful
eyewitness, verifying for American TV viewers that U.S. bombs were exploding caves near Tora Bora and destroying what he called the "rats' nest."

Just as *Our World* brought several satellite television practices to the fore, initiating the central questions of this book, so too did 9/11 coverage seem to bring them to a close, intertwining as it did many of the satellite television forms and modalities I have discussed. However problematic its claims of first world excellence and third world trouble, the liberal human-ism and global village utopianism of *Our World* was altogether transformed and recoded after 9/11. The most recent forms of satellite television incar-nate and valorize a frightening blend of military monitoring and commer-cial entertainment.

After the decisive blow to American soil, live satellite transmissions, direct satellite broadcasting, and remote sensing have been used not to unify the world but increasingly to divide and patrol it. Satellite broadcast-ing services, whether CNN, Fox News, or Al-Jazeera, have heightened and exaggerated differences between the Muslim East and Christian West, the subterranean terrorists and the Texan good old boys, the desolate Afghani and Iraqi deserts and bustling American cities. In addition, satellite images have appeared in TV news with unprecedented frequency. Since 9/11 they have exposed bombed Al Qaeda targets in Afghanistan, alleged nuclear weapons facilities in North Korea, and UN weapons treaty violations in Iraq. Almost every threat to U.S. security, it seems, can be discovered via satellite except Osama bin Laden himself.

The hard lines drawn by satellite television's planetary patrols coincide with nationalistic legal and social acts as well—acts ranging from the Pa-triot Act to American flag waving, from the creation of a Homeland Secu-rity Agency to the Immigration and Naturalization Service's mass interro-gation and expulsion of South Asian and Arab men. Despite the potential of satellite television to represent the world in culturally diverse and com-plex ways, we live in an age in which the U.S. president paints the planet as two spheres of "good and evil" and audaciously declares to the global community, "You're either with us or against us!"

And so, the collapse of the towers became TV networks' most beloved in-stant replay. And satellite images of ground zero became a splendid open wound. But rather than embrace these forms of satellite television as 9/11 "truths," I found myself deeply skeptical once again. My trip to Bosnia earlier that year had confirmed that as much as we may use the terrains of satellite television to determine our global positions, these grounds are themselves always shifting, indeterminate, and uncertain. This recognition of satellite television's precarious relation to truth is vital since dominant institutions and ideologies have historically hailed it as tantamount to what Foucault calls the "perfect historian." In *The Archaeology of Knowledge* he writes, "The [perfect] 'historian' desires total knowledge, avoids the exceptional and reduces it to the lowest common denominator and he dominates those of the present with his pretensions of knowledge and perspective; he is demagogic but this demagogry is masked under the cloak of universals. Divided against himself (he is also that which is in the particular), he discloses the (mystifying) eternal will, rejecting his subjectivity for a 'higher' objectivity."[11] Satellite television technologies have been used again and again to affirm universals, to offer the ultimate views, the most authoritative forms of knowledge, and the highest objectivity. Yet as I suggested at the outset of this book, the satellite's physical displacement from Earth also positions it as an alienated objectivity divided against itself.[12]

Both remote sensing and astronomical observation can generate alienated objective views that symbolically displace the earth from its conceptual center in the universe. These satellite television practices have rendered our planet, on one hand, as a digital abstraction requiring close analysis to make any sense, and, on the other hand, as a tiny ball in an infinitely expanding domain of outer space, always on the verge of disappearance. Satellite television practices can be used both to affirm and undermine Enlightenment paradigms that are organized by distant vision, knowledge, and universality. They can both advance and complicate discourses of Western Eurocentrism.

This is why Foucault's concept of genealogy is ultimately more appropriate for this study of satellite television than is the notion of "perfect history." For genealogy emphasizes discontinuities, ambivalences, and contradictions as a way of destabilizing dominant power formations. In the process of establishing satellite television's relation to the global, this genealogy has attempted to stir instability within claims that the satellite is science's greatest technology of objectivity, humanity's most thorough historian, and militias' most strategic weapon.

"Situated within the articulation of the body and history," genealogy, according to Foucault, "shortens vision to those things nearest it—the body, the nervous system, nutrition, digestion and energies."[13] In an effort to lessen the gap between the satellite's orbit and the earth, I have explored how satellite television practices have produced and imprinted themselves on bodily surfaces. For instance, *Our World*'s live signal compared the massified, hungry, and over-reproducing body of third world countries to a first world individualized body of athletic and cultural excellence. The U.S. satellite images of Bosnian mass graves submerged bodies that matter in a field of digital abstraction. Archaeologists used satellite television to re-vivify and make contact with the body of Cleopatra. Hubble views of deep space honed in on human reproduction as the site of stellar formation. And satellite television coverage of 9/11 brought scars on the national body into bold relief. Because of the satellite's distance from the earth and its displacement of human vision, the discourses generated by and surrounding its uses often return to the contours of the flesh. For it is only through the body and the senses that we can ultimately understand satellite technologies and their global effects.

Pushing this logic further, we might consider how satellite television, genealogy, and the body come together to form a critical practice of global positioning. So far this discourse has been manifest obliquely as a series of longitudinal and latitudinal coordinates opening each chapter. As I have discussed elsewhere, global positioning is a satellite-based system of mapping and orientation developed for the U.S. military in the 1980s, but it can function as a technology of social positionality as well.[14] What I am proposing here, then, is much more than the capacity to orient one's self within a global numerical grid. Global positioning is, I think, a very useful metaphor for the contemporary study of media technologies in general and particularly for those that draw on and combine social constructivist and genealogical approaches.

We might imagine "global positioning" as an ongoing critical appropriation (and in some cases, reversal) of the U.S. military's technological paradigm of "anytime and anywhere" to promote a practice of global partiality. The term *global positioning* implies the need to formulate ways of describing and analyzing the socioeconomic forces that constitute global telecommunication, television, and computer networks as well as the audiovisual signals distributed through them. Network technologies are so decentralized that they often seem beyond critical reach.[15] To study technologies of such extensive scale and high speed inevitably requires some act of global posi-

tioning. To identify such technologies as objects of analysis, scholars must
select and situate their work within sociohistorical conditions, geographic
territories, discursive sites, textual forms, and industries.

This practice of positioning need not be limited to the study of technology in local or national cultures. As David Morley argues, "given the extent of mobility of both populations and media flows across national boundaries in the contemporary world, any model of culture and communications that operates solely within the assumptions of a national framework is inadequate."[16] The challenge, it seems to me, is to find ways to work transnationally while differentiating the analysis of global technologies (the "global positioning," if you will) from the imperializing undertones of travel, which, as Caren Kaplan so eloquently explicates, shape both scholarly practice and world conditions.[17] Put more simply, how can we conduct analyses of global technologies and not be imperialists?[18] Perhaps assuming up front that technological analysis requires acts of global positioning could be a start. For we can only ever access and understand the uneven effects of global technologies such as satellite television or the World Wide Web through transnational and comparative forms of analysis that isolate and engage specific yet disparate instances of use. As Thomas Elsaesser suggests, we need "a more conjectural history of the technical media, respecting uneven developments and discontinuities."[19] Metaphorically, global positioning involves clutching parts of an unwieldy machine and tinkering with them to create a telling amalgam of the machine rather than a fully molded copy. It is a practice of critical tinkering.

In this sense global positioning functions in a way similar to genealogy as well. It requires the critic to plot courses of technological intelligibility rather than a linear path toward total comprehension of a perfect machine. It involves an attempt to expose how technologies and knowledge practices circumscribe and produce world geographies, economies, and cultures. It plunges into the dialogic relations between technological uses, knowledge practices, and social positionalities that stretch across national borders. It requires understanding of how satellite television practices, for example, become discursive sites through which viewers/subjects/citizens both make sense of and stake out positions in the world, positions that are, of course, never fixed but complex and constantly renegotiated.

Just as I am calling for a critique that does not sidestep the global proportions and effects of technological uses, I believe there is a need for further consideration of the disparate practices of satellite television reception and viewing as well. Shaun Moores, Purnima Mankekar, and others have conducted pathbreaking studies of direct satellite broadcasting audiences in England and India respectively.[20] While it is crucial to consider how DBS impacts national and local cultures in different parts of the world, it is equally important to explore how other satellite television practices such as live international transmission, remote sensing, and astronomical observation impact viewing and knowing as well. How did viewers outside of the United States use live satellite transmissions to interpret the 9/11 attacks? How do viewers engage on a daily basis with satellite images of weather? How do viewers of astronomical images use them to confirm their religious beliefs? How are the analysts of military satellite intelligence trained to see? By thinking across borders, through historical periods, across different uses of the technology, the practice of global positioning is intended to activate models of analysis that are nonnationalist and interdisciplinary.

The critical tendency to associate satellite television exclusively with dominant institutions and ideologies is itself dangerous and politically counterproductive, for it has the effect of triggering an implicit surrender in the social struggle over future uses of the technology. If I were to isolate one important precept from the social constructivist approach to the study of technology, it would be that technologies are processes, not things, and as such their forms can change and their uses can be redirected. This means that all technological uses are potential sites of social and political intervention. Since satellite television practices have recently been harnessed to help foster what some critics have described as a New World Order, it is vital, as Foucault reminds us, that we continue to "place the mechanism that establishes new truths under strict supervision."[21] It is vital, in other words, that we continue to study, critique, and intervene in practices of satellite television since they are being mobilized to intensify and reinforce global militarization.

On the cover of his book *Feeling Global* Bruce Robbins reprints several aerial photographs shot by his father during World War II to convey a structure of feeling that he calls "internationalism in distress." As he explains, "It cannot be accidental that I associate worldliness with world war, the desire to know with the view through a bombsight, and wider horizon with

the altitude of an aircraft. . . . The bombsight perspective that seems to
threaten also provokes a longing to overcome these distances."[22] As I have
suggested throughout this book, satellite television is necessarily implicated within similar international distresses, for we cannot talk about the
fantasy of a global village without considering the hard edge of geopolitics
that satellites carved and maintained during and after the cold war. We cannot imagine the euphoria of transnational connections without recognizing the many peoples who have historically been excluded from them. We
cannot look at a colorful view of deep space without recognizing that spy
satellites are turned the other way, monitoring activity in our backyards.

As much as satellite technologies have enabled us to see beyond the
earth, they have also positioned the planetary practices as objects-to-be-
looked-at. As Thierry Jutel suggests, "I can't help thinking that going into
outer space is also searching for a reverse shot, for a position from where
all systems of representation would be brought to a closure so that the
demand/need/desire for the duality of the image—point of view/object—
could be replaced by the paradoxical emergence of the cosmos as image
without a point of view."[23] The satellite does not so much collapse the
duality of perspective as it amplifies the possibility of alienation or difference within the field of the televisual. This amplification occurs, for instance, in the haunting possibility that events on the earth could always be
seen and encoded from anOther (nonhuman) point of view. It also occurs
by virtue of the satellite's own limitations in representing events on the
earth. The world is, of course, experienced all the time in unseen, unknown, and untold ways.

If, as Paul Virilio suggests, the glow around the earth is the last stage in
history, then satellites help shape the contours of this luminous fog.[24] This
is the domain Peljhan calls the "signal territory"—a space born of intricate
paths drawn through the ether. While the earth's afterglow may be mag-
nificent, as the last stage in history it is certainly not innocent. In this fuzzy
twilight of orbiting satellites, signal crossings, and flying communication,
the past, present, and future loom, and practices of culture, science, and
militarism collide. I began this act of global positioning with my eyes transfixed on one passing satellite, but I end it with my critical gaze turned on
the nebulous realm of the signal territory. For it is in this obscure haze that
satellite-based transactions transpire. Throughout this book I have traced

discursive threads that tether these technologies to life on the earth and to our ongoing attempts to make sense of our world through the televisual. But more global positioning will be necessary before we can truly appreciate the complex trajectories of satellite television and the traces left in their wake.

Notes

Introduction

1 By *discursive modality* I mean a form of address that is associated with particular knowledge practices.
2 Jameson, preface, p. xvi.
3 Stucker, "Junkosphere," p. 40.
4 In 1994 Chicago Chinese Communications, Inc., distributed Chinese language programming via satellite to more than five hundred thousand households in North America. See "The Chinese Connection."
5 Davan Maharaj, "How Tiny Qatar Jars Arab Media," *Los Angeles Times*, May 7, 2001.
6 Fair, "Francophonie and the National Airwaves."
7 "Africa Online and Teleglobe Launch Internet Satellite Link for Ghana; Teleglobe to Also Serve Cote d'Ivoire and Kenya." For a discussion of indigenous uses of satellite broadcasting see Batty, "Singing the Electric." See also Stenbaeck, "Communication, Culture and Technology."
8 The policy on Foreign Access to Remote Sensing Space Capabilities allowed American companies to sell satellite images up to one meter in resolution. The Department of Commerce issues licenses, and the Department of Defense supervises and monitors licensing to protect national security interests. See "Statement on Export of Satellite Imagery and Imaging Systems."
9 The first weather satellite, *Tiros*, was launched on April 1, 1960, and was equipped with two cameras. Its images were immediately used by weather forecasters and televised (see "Tiros-1").
10 Anselmo, "Remote Sensing to Alter TV News," p. 61.
11 For further information about Dennis Tito's space tour see "First Space Tourist." For a discussion of Space Marketing's plans see "Introduction of Legislation Regarding Billboards in Space," *Congressional Record*, July 1, 1993, pp. E1732–E1734.
12 Giddens, *Beyond Left and Right*, p. 80.
13 Baudrillard, *Simulations*, p. 63.
14 Ibid, p. 62. Baudrillard claims that models of security are encouraging "satellisation of the whole planet. . . . For what is the ultimate function of the space race, of lunar conquest, of satellite launchings, if not the institution of a model of universal gravitation, of satellisation, whose perfect embryo is the lunar module: a programmed microcosm, where *nothing can be left to chance*?" (p. 62). He also writes, "With satellisation, the one who is satellised is not whom you might

think. By the orbital inscription of a space object, the planet earth becomes a satellite, the terrestrial principle of reality becomes excentric, hyperreal and insignificant. By the orbital establishment of a system of control like peaceful coexistence, all terrestrial microsystems are satellised and lose their autonomy" (p. 64).

15 Moores, "Satellite TV as Cultural Sign," p. 636.

16 Ibid., p. 623. Moore's work critiques satellite television in the context of neighborhood affinity, consumer desires, and social mobility.

17 Brunsdon, "Satellite Dishes and the Landscapes of Taste," pp. 23–37.

18 Berland, "Mapping Space," p. 124.

19 Ibid., p. 128.

20 Williams, *Television*, p. 44.

21 As Williams writes, "technology is at once an intention and an effect of a particular social order" (ibid., p. 128).

22 Ibid., p. 143.

23 Ibid., p. 144.

24 See Williams, *Towards 2000*. Also see Francis Mulhern, "*Towards 2000*, or News from You-Know-Where."

25 Feenberg, *Questioning Technology*, p. x. For an earlier description of the social constructivist approach see MacKenzie and Wajcman, *The Social Shaping of Technology*.

26 Feenberg, *Questioning Technology*, p. 11.

27 As Williams writes, "Unlike all previous communications technologies, radio and television were systems primarily devised for transmission and reception as abstract processes, with little or no definition of preceding content. When the question of content was raised, it was resolved, in the main, parasitically. There were state occasions, public sporting events, theaters, and so on, which would be communicatively distributed by these new technical means. It is not only that the supply of broadcasting facilities preceded the demand; it is that the means of communication preceded their content" (*Television*, p. 25).

28 Williams described "visual mobility" as the "ability to watch things as various . . . as horse-racing, a street interview, an open-air episode of a play or a documentary" (*Television*, p. 77). For another discussion of visual mobility see Friedberg, *Window Shopping*.

29 Caldwell, *Televisuality*, p. 5.

30 Introduction to Reid and Traweek, *Doing Science + Culture*, p. 7.

31 For other work that considers television in public space see McCarthy, *Ambient Television*; and Couldry and McCarthy, *Media Space*.

32 Williams, *Television*, p. 135.

33 One of the exceptions to this has been the emergence of the Deep Dish Television Network, which used satellite time to distribute activist videos and television programs by groups such as Paper Tiger TV.

34 As Katherine Hayles explains, the "sciences of complexity" have their origins in chaos theory. For further discussion see her article, "Narratives of Artificial Life."

35 See, e.g., Keller and Grontowsky, "The Mind's Eye," p. 207; and Keller, *A Feel-*

ing for the Organism. Also see discussion of these works in Braidotti, *Nomadic Subjects,* pp. 71–73. For further discussion of these issues see Harding, *Whose Science?;* and Treichler et al., *The Visible Woman.*

36   Braidotti, *Nomadic Subjects,* p. 71.

37   Haraway, *Simians, Cyborgs, and Women,* p. 188.

38   Ibid., pp. 191–192. As Haraway writes, we "don't want to theorize the world, much less act within it, in terms of Global Systems, but we do need an earth-wide network of connections, including the ability partially to translate knowledges among very different—and power-differentiated—communities" (p. 187).

39   Foucault describes a "history that would be not division, but development (*devenir*); not an interplay of relations, but an internal dynamic; not a system, but the hard work of freedom; not form, but the unceasing effort of a consciousness turned upon itself, trying to grasp itself in its deepest conditions: a history that would be both an act of long, uninterrupted patience and the vivacity of a movement, which in the end, breaks all bounds" (Foucault, *The Archaeology of Knowledge,* p. 13).

40   For critical discussion of "the West" see Shohat and Stam, *Unthinking Eurocentrism,* pp. 13–15.

41   See John Beaufort, "Girdling the Globe Electronically," *Christian Science Monitor,* June 26, 1967.

Chapter 1. Satellite Spectacular:
Our World and the Fantasy of Global Presence

1   *Our World* press packet, May, 1967, NET Collection, National Public Broadcasting Archives, Hornbake Library, University of Maryland, College Park, Md.

2   See John Beaufort, "Girdling the Globe Electronically," *Christian Science Monitor,* June 26, 1967; Delatiner, "Global Show Displays Electronic Magic"; Mary Ann Lee, "'Our World' Sees Advent of New Eras in Television." The original title of the show was *Round the World in 80 Minutes,* evoking Jules Verne's prophecy, and then it became *Spaceship Earth—a View of Man on the Planet.* Other suggestions included *The World Is Ours* and *Without Horizons.*

3   Caldwell, *Televisuality,* p. 252.

4   Sconce, *Haunted Media,* p. 6.

5   For further discussion of modernization theory in the 1950s and 1960s see Harrison, *The Sociology of Modernization and Development,* pp. 29–31.

6   Coyle, "TV's Global Future Awaits Those Who Take the Opportunity Today," p. 93.

7   "Telstar's TV Future Is Foggy," p. 33.

8   Countries that participated in the planning of the show included Tunisia, Japan, Austria, France, Italy, Sweden, Spain, England, West Germany, East Germany, the Soviet Union, Czechoslovakia, Hungary, Poland, Canada, Mexico, the United States, and Australia. See Potter, "Memo to All Stations."

9   "'Our World'—Round the World Project Fact Book."

10   For further discussion of the treaty see Luther, *The United States and the Direct Broadcast Satellite,* pp. 68–69.

11 UNESCO, *Communication in the Space Age*, p. i. For further discussion and cri-
    tique of Western discourses on global television during the 1960s see Curtin,
    Redeeming the Wasteland.

12 UNESCO, *Communication in the Space Age*, p. 20.

13 Ibid., p. 43.

14 Ibid., p. 38.

15 Ibid., p. 145.

16 Ibid., pp. 120–121.

17 Ibid., pp. 117–118.

18 Cited in Luther, *The United States and the Direct Broadcast Satellite*, pp. 82–83.

19 The preamble to the INTELSAT agreement expressed the members' desire to
    establish a "single global commercial communications satellite system . . . which
    will contribute to world peace and understanding" (cited in Luther, *The United
    States and the Direct Broadcast Satellite*, p. 82).

20 Levy, "Memo to All Stations."

21 See "Conflict Splits World Telecast." The Six-Day War broke out on June 5, 1967,
    when Egypt concentrated forces in the Sinai Peninsula and conspired with Jor-
    dan, Syria, Algeria, Kuwait, and other Arabic states to threaten Israel's borders.

22 "Round the World Project."

23 Lenica and Sauvy, *The Population Explosion*; Hauser, *The Population Dilemma*;
    Appleman, *The Silent Explosion*; and Ehrlich, *The Population Bomb*.

24 Ehrlich, *The Population Bomb*, p. 17.

25 Western scientists and intellectuals imagined various ways of resolving the
    population problem, ranging from family planning to sending people to the
    moon and other planets in the solar system. Other alternatives included mass
    use of antifertility agents in the water or staple foods, temporary sterilization of
    all girls at puberty, compulsory abortions of illegitimate pregnancies, and cash
    bonuses for men who get vasectomies. See "Interstellar Migration and the Popu-
    lation Problem"; and Notestein, Kirk, and Segal, "The Problem of Population
    Control," p. 164.

26 Erlich, *The Population Bomb*, p. 22.

27 In this sense *Our World*'s globalist discourse resonated with earlier cultural
    forms such as Edward Steichen's 1955 traveling photographic exhibition *The
    Family of Man*, which, as Steichen explains, was "conceived as a mirror of the
    universal elements and emotions in the everydayness of life — as a mirror of the
    essential oneness of mankind throughout the world" (Steichen, introduction to
    The Family of Man, p. 3).

28 McLuhan and Fiore, *The Medium Is the Message*, p. 63.

29 In this way *Our World* also resembled the NBC television series *Wide Wide World*
    (1954–1957), which was a biweekly, ninety-minute "super-spectacular" featur-
    ing live transmissions from locations across North and Central America. Like
    Our World, the show addressed viewers as "armchair travelers" and emphasized
    themes of global mobility and "worldliness." For further discussion see Parks
    "As the Earth Spins."

30 *Our World* script.

31 Ibid.

32 "'Our World'—Round the World Project Fact Book."

33 Wilson, press release regarding "Our World."

34 *Our World* script.

35 Lee, "'Our World' Sees Advent of New Eras in Television."

36 Wilson, press release regarding "Our World."

37 Ibid.

38 "'Our World'—Round the World Project Fact Book."

39 Elsaesser, "Digital Cinema," p. 210.

40 One review described the broadcast as "an old fashioned geography class gone electric." See Lee, "'Our World' Sees Advent of New Eras in Television."

41 *Our World* press release.

42 *Our World*'s "Production Philosophy," p. 6.

43 Colin McCabe first develops the concept of a hierarchy of discourse in his essay "Realism and the Cinema."

44 For a discussion of liveness in 1980s television see Jane Feuer's classic essay, "The Concept of Live Television." For an interesting discussion of electronic presence in 1960s television see Jeffrey Sconce's "The 'Outer Limits' of Oblivion."

45 Producers suggested, "Whenever possible individual remote contributions should be constructed so as to develop through time rather than across space; the metaphor in the producers [sic] mind should be that of lighting a fuse rather than of going around a garden gathering flowers. We should pick a single operation and watch it develop rather than looking at several which do not develop at all" ("'Our World'—Round the World Project Fact Book").

46 Appadurai, *Modernity at Large*, p. 3.

47 Ibid., p. 9.

48 "Photo Cutlines."

49 *Our World* script.

50 Ibid., p. 28.

51 For a discussion of the politics of reproduction in third world countries see Morsy, "Biotechnology and the Taming of Women's Bodies," pp. 165–173.

52 Decades later global satellite providers would aggressively target audiences and exploit markets in precisely those countries that were allegedly the primary culprits of the population explosion, including India, China, Brazil, and South Africa.

53 See Crary, *Techniques of the Observer*.

54 This flurry of memos suggests producers were undergoing some anxiety about characterizing the show in this manner. See, e.g., Lou Potter, "Memo to All Stations."

55 See "Babies Are Focal Point."

56 Bob Tweedell, "'Our World' Impressive," *Denver Post*, June 26, 1967.

57 Related to *Our World* are more-recent projects such as "The World Right Now," a Web site that "gathers virtually live outdoor views from video cameras set up in 148 locations around the globe," including "a street in central Karachi, the

Wailing Wall in Jerusalem, a faceless intersection in Memphis, the Los Angeles downtown skyline." This site is discussed in Joe Sharkey, "Step Right Up and See Grass Grow and Paint Dry," *New York Times*, March 22, 1998, p. 4.

58 Caldwell, *Televisuality*, p. 31.

59 Sassen, "Saskia Sassen on the Twenty-First-Century City." Also see Sassen, *The Global City*; and Sassen, "The Global City: Strategic Site/New Frontier."

Chapter 2. Satellite Footprints:
Imparja TV and Postcolonial Flows in Australia

1 Contractor et al., "Metatheoretical Perspectives on Satellite Television and Development in India"; McAnany and Oliveira, *The SACI/EXERN Project in Brazil*; and Pelton, "Project SHARE and the Development of Global Communications." See also Parapak, "The Roles of Satellite Communications in National Development"; and Javier Esteinou Madrid, "The Morelos Satellite System and Its Impact on Mexican Society."

2 Imparja's shareholders include Aboriginal groups from central, northern, and southern Australia. They include the Central Australia Aboriginal Media Association, Central Land Council, Tiwi Land Council, Warlpiri Media Association, Pitjantjatjara Council, Top End Aboriginal Bush Broadcasting Association, Maralinga Tjaritja Trust, and the Aboriginal and Torres Strait Islander Commission. Many of these groups made money from business holdings and by leasing mining rights on their land.

3 Michaels, *The Aboriginal Invention of Television, 1982–1986*. For a discussion of this work see O'Regan's essay "TV as Cultural Technology."

4 Dirlik, "The Postcolonial Aura," p. 503.

5 In 1998 Imparja had a potential audience of 189,000 residents in areas throughout Northern Territory, South Australia, Victoria, and New South Wales. See Imparja Television Web site.

6 In her very insightful work on Aboriginal media, Faye Ginsburg offers a contextual model for studying such issues as she refuses to separate "textual production and circulation from broader arenas of cultural production" (p. 368). She suggests, "In the imaginative, narrative, social and political spaces opened up by film, video, and television are possibilities for Aboriginal media makers and their communities to re-envision their current realities and possible futures" (Ginsburg, "Embedded Aesthetics," p. 382).

7 Williams writes, "What is being offered is not . . . a programme of discrete units with particular insertions, but a planned flow, in which the true series is not the published sequence of program items but this sequence transformed by the inclusion of another kind of sequence, so that these sequences together compose the real flow, the real broadcasting" (Williams, *Television*, p. 84).

8 White, "Flow and Other Close Encounters with Television."

9 For a discussion of flow see Williams, *Television*, pp. 89–93. Also see Corner's *Critical Ideas in Television Studies*, pp. 60–69.

10 See Cocca. "Consent, Content, Spillover and Participation in Direct Broadcasting from Satellites."

11 "What Is Imparja Television?"

12 Quoted in Batty, "Singing the Electric," p. 110.

13 Ibid., p. 124.

14 This program was the result of a federal government report entitled *Out of the Silent Land*, commissioned by the Dept. of Aboriginal Affairs in 1985 to evaluate Aboriginal broadcasting.

15 Langton, *"Well, I Heard It on the Radio and I Saw It on the Television,"* p. 17.

16 "The Black TV Moguls with Carte Blanche in the Outback."

17 Ibid.

18 Ibid.

19 United Press International, "Australian Aborigines Get TV Station License," *The Record*, Sept. 28, 1986, p. A44.

20 "Australia's First Aboriginal TV Opens."

21 *"Dallas* in the Outback."

22 As Katz and Liebes have demonstrated in their ethnographic studies of *Dallas* reception, communities responded differently to the program. Katz and Liebes examine interpretations of *Dallas* by four ethnic communities in Israel (Arab, Russian, Moroccan, and Kibbutzim). See Liebes and Katz, "On the Critical Abilities of Television."

23 See Batty, "Singing the Electric."

24 Michaels, in *Bad Aboriginal Art*, writes, "The simple logistics of providing for all these languages on a single service clearly indicates a fundamental mismatch. The bias of mass broadcasting is concentration and unification; the bias of Aboriginal culture is diversity and autonomy. Electronic media are everywhere; Aboriginal culture is local and land-based. Only local communities can express and maintain linguistic autonomy" (p. 100).

25 A television project called "Footprints" was also released in 1993 in conjunction with the International Year of the World's Indigenous Peoples. The series was produced by Sorena Productions, and it featured 105 short public service segments on Aboriginal culture and history that were interspersed between other regularly scheduled programming. One segment, called "Landcare," discussed Central Australian Aboriginal Land Councils' pooling of talents to make an award-winning documentary promoting care of the land. Tanami Network discussed video satellite networks that link remote desert communities, "allowing them to talk to each other and the outside world." Others feature Mum Shul, a well-known Aboriginal social worker; famous tennis player Evonne Goolagong; and an Aboriginal cricket team's trip to England. See "FOOTPRINTS — Storylines."

26 "Alice Springs Television Production Company Embarks on a Brave Project," *Alice Springs News*, June 11, 1997.

27 "CAAMA Productions."

28 Corallie Ferguson, Chief Executive Officer, Imparja Television, to Lisa Parks, FAX, Sept. 23, 1997.

29 For an interesting discussion of the Dreaming as it appears in Eric Michaels's work see Hodge, "Aboriginal Truth and White Media."

30 Arden, *Dreamkeepers*, pp. 3–4.

31 Michaels also explored how Aborigines brought the new means of video and television "inside the Law." He claims that to understand Warlpiri video, "one must appreciate this sense of Jukurrpa as a model of social reproduction—and identify the ways in which novelty must be counterinterpolated to the system." *Jukurrpa* is divided into secret and public domains, partitioning information and restricting access. Determination of secrecy is not absolute, however, and it can sometimes be negotiated locally. As Michaels explains, "Warlpiri law restricts the movement of information through social and geographic space, encouraging it to accrue value to itself. Consequently, secret knowledge can become the most valued item of exchange in Warlpiri economy. People who compromise this system today are still scorned for 'selling their law'" (Michaels, *Bad Aboriginal Art*, pp. 107–108).

32 Ibid.

33 Beatson, "Dreamtime Beam."

34 Langton, *"Well, I Heard It on the Radio and I Saw It on the Television,"* p. 19.

35 For further discussion of the representation of Aboriginals in television and film see Langton, *"Well, I Heard It on the Radio and I Saw It on the Television"*; Malone, *In Black and White and Colour*; Jennings, *Sites of Difference*; Leigh, "Curiouser and Curiouser."

36 Lola Forrester, radio interview by Andrew Coulloupas, "The Media and Aboriginal Reconciliation."

37 Ferguson to Parks.

38 "Yamba." The honey ant is the network's mascot and travels throughout the central zone footprint to visit schools and community centers.

39 "Maintaining Aboriginality at Imparja."

40 As Imparja's news manager Bruce Honeywell explains, "We try to get the stories that people can identify with in the small communities, which is quite apart from mainstream media. . . . Our approach is different; we actually let people finish sentences and things like that" (ibid.).

41 Lowe, *Immigrant Acts*, p. 67.

42 Newcomb and Hirsch, "Television as a Cultural Forum," p. 509.

43 Langton, *"Well, I Heard It on the Radio and I Saw It on the Television,"* p. 17.

44 Batty, "Singing the Electric," p. 124.

45 "Imparja: Keeping Strong," 1998, p 3.

46 "High Tech for the Bush Aborigines."

47 Quoted in O'Regan, *Australian Television Culture*, p. 190.

48 In August 2001 the most popular programs were *ER*, *Friends*, an Australian series called *Blue Heelers*, *Big Brother*, *Popstars*, *Survivor*, travel shows such as *Getaway*, and *Who Wants to Be a Millionaire*.

49 In *Satellite Dreaming* Clive Scollay says in those communities that do not actively use BRACS equipment, the satellite dish has become a "sponge that just sucks up what the satellite has to offer."

50 Kiernan Finnane, "Aboriginal Imparja Is Television with Fire in Belly," *Alice Springs News*, June 18, 1997.

51 Michel de Certeau suggests that consumption has become a significant form of cultural production. See de Certeau, *The Practice of Everyday Life*.

52 Naidoc Week promotional pamphlet, Aboriginal and Torres Strait Islander Commission, July 1999.

53 Donna Campbell, interview by author, Imparja TV offices, Alice Springs, Australia, July 14, 1999.

54 Much of this music is itself considered hybrid. In his analysis of Aboriginal pop music by Yothu Yindi and the Warumpi Band, Tony Mitchell suggests, "Historically, Aboriginal popular music in Australia has tended to reflect prevalent white Anglo-American influences of country and western rock, as well as reggae, which are now being combined with traditional Aboriginal musical forms and new dance music technologies in a hybridization similar to, but also significantly different from, the black musics in Britain described by [Paul] Gilroy" (Mitchell, "Treaty Now! Indigenous Music and Music Television in Australia," p. 301). Many of their music videos draw on the conventions of MTV while using local settings, actors, and sounds.

55 Jacqui Bethel, interview by author, CAAMA offices, Alice Springs Australia, July 16, 1999.

56 *Dawson's Creek* Web site.

57 A segment in *Satellite Dreaming*, for instance, focuses on the production of an educational video in the village of Yuendumu. CAAMA producers are shown working with Warlpiri actors to create a video that will teach children to count in their native Warlpiri language. The segment borrows as much from Western pop music videos as it does from Aboriginal dreaming stories, as it combines catchy music, eclectic montage, and air guitar sequences. In the segment a white teacher in that community stresses the importance for children to see themselves on television and to have programs they can call their own. Such productions are especially important given that when popular Aboriginal bands have appeared in global forums such as the MTV awards broadcast, their political iconography and/or land rights claims are often submerged or removed. See Mitchell, "Treaty Now! Indigenous Music and Music Television in Australia," p. 305.

58 Donna Campbell, interview with author, July 14, 1999, Imparja TV offices, Alice Springs, Australia.

59 Parry, "Problems in Current Theories of Colonial Discourse," p. 42.

60 Kieran Finnane, "CAAMA to Produce Outback Soapie," *Alice Springs News*, June 17, 1998.

61 Browne, "Aboriginal Radio in Australia."

62 "Television Goes Global," p. 95.

63 Beatson, "Dreamtime Beam."

64 Ernabella is a remote community in northern South Australia. Through the Pitjantjatjara Land Rights Act, the Pitjantjatjara people won control over their traditional lands, which span an area half the size of England. Neil Turner, media co-

ordinator for EVTV says, "The custodians [of dreaming, or *Tjukurrpa*] have found video a particularly apt tool for recording the unique and integral combination of story, song, dance and visual arts and landscape through which the Tjukurrpa is expressed . . . as it can be entirely directed, produced, and distributed by themselves, without the expertise of literate white anthropologists" (quoted in Batty, "Singing the Electric," p. 112).

65 Michaels, *Bad Aboriginal Art*, p. 110.

66 Ibid., p. 114. Also see Ginsburg, "Embedded Aesthetics."

67 Another Yuendumu video represents the fire ceremony "Warklukurlangu," which had seldom been performed since ethnographers Roger Sandall and Nicolas Peterson filmed it in 1967. "The fire ceremony," Michaels explains, "was an explicit expression of Warlpiri autonomy, and for nearly a generation it was obscured" (ibid., p. 116). The presence of the video camera brought back some dances and local rituals that had not been performed for a long time. The tape of the fire ceremony became very popular in Warlpiri communities, and it became difficult to keep track of the copies. When one of the central performers died shortly thereafter, the tapes were collected, put on the shelf, and labeled "not to look," since it is against Warlpiri tradition to look at a photograph of a person once he or she has died (ibid., p. 118). As Michaels explains, the representation of some things in media or photography is forbidden (representations of dead people, for example; and a man cannot look at his mother-in-law if she appears full-face on the TV screen because to do so would compromise traditional rules that govern this relationship). It was controversial for Michaels to take the videos out of the Yuendumu archive and discuss them in his book. Some Aborigines, however, believe that there is something to be gained by discussing their art in front of the world. Although the "Warklukurlangu" videotape was initially part of a cultural revival, it ultimately became part of an archive that could not be reproduced; in effect, the videotape died when one of the performers died. The challenge for the Warlpiri, then, is to use video and satellite media to maintain local dreamtime knowledges within the bodies of the living—for the knowledges of those who die, even though captured on videotape, cannot necessarily be reproduced. This was confirmed by my interview with Jacqui Bethel, director of CAAMA, July 20, 1999, CAAMA offices, Alice Springs, Australia.

68 Hamid Naficy discusses his book *The Making of Exile Cultures: Iranian Television in Los Angeles* (Minneapolis: University of Minnesota Press, 1993) in his essay "Between Rocks and Hard Places: The Interstitial Mode of Production in Exhilic Cinema," in Naficy, *Home Exile, Homeland*, 135.

69 This use of lowcasting by Aboriginal Australians might be compared to Black Liberation Radio in the United States, a system by which African Americans use low-powered "pirate" transmitters to broadcast radio programs to highly local communities. Such programs provide an important alternative to white mainstream media, and they can be a vital source of local solidarity and community building. The "blackstream counter-knowledges," of Black Liberation Radio, John Fiske claims, "can provide a deconstructive jolt" to "official knowledge"

(Fiske, *Media Matters*, p. 192). Similarly, some Aboriginal communities have seized on the technical resources of BRACS equipment and generated media that serve the cultural needs of very small local audiences.

70 See paper by chief engineer, Tim Mason, "Conversion of Imparja's Remote Area Satellite Television Services to Digital," delivered to me by the author July 1999.

71 Ruth Lamb-Carlson, director of commercial sales, interview by author, Imparja TV offices, Alice Springs, Australia, July 15, 1999.

72 "National Indigenous Television Service Begins Test Transmissions," Imparja media release, Aug. 9, 1999.

73 Barry Clarke, anchor *Imparja National News*, interview by author, Imparja TV offices, Alice Springs, Australia, July 14, 1999.

74 Dwayne Tickner, interview by author, Imparja TV offices, Alice Springs, Australia, July 12, 1999.

75 Batchen "Da(r)ta," p. 6.

76 Cited in Buchtmann, "Digital Songlines," p. 8

77 Hodge, "Aboriginal Truth and White Media," p. 11.

78 Batty, quoted in the *Satellite Dreaming* documentary.

79 "Aboriginal Map Acquisitions."

80 Batty, "Singing the Electric," p. 107.

81 Menser and Aronowitz, "On Cultural Studies, Science, and Technology," p. 14.

82 Canclini, *Hybrid Cultures*, p. 190.

Chapter 3. Satellite Witnessing:
Views and Coverage of the War in Bosnia

1 Berland, "Mapping Space," p. 128.

2 Menser and Aronowitz, "On Cultural Studies, Science, and Technology," p. 15.

3 The *Corona* project was declassified in January 1996 and has since become the subject of several television documentaries, such as the Discovery Channel's *Eye in the Sky* (March 10, 1996) and *Spy Watch* (March 11, 1996). For further discussion of the *Corona* project see Day et al., *Eye in the Sky*; and Brown, "America's First Eyes in Space."

4 The *Corona* project's satellite images were acknowledged publicly on February 23, 1995, when President Clinton declassified them. See "Declassified Intelligence Satellite Photographs." *Corona*'s first image was of a Soviet airfield. Others showed nuclear weapons facilities, long-range airfields, Soviet aircraft, missile manufacturing plants, and lunar launch facilities.

5 Mack, *Viewing the Earth*, p. 3.

6 Curtis, *Space Satellite Handbook*, p. 83.

7 McElroy, "Observing the Earth from Space," p. 20.

8 William J. Broad, "Private Ventures Hope for Profits on Spy Satellites," *New York Times*, Feb. 10, 1997, p. 1.

9 One of the most symbolic uses of the remote-sensing satellite in recent times has been Vice President Al Gore's $25–50 million proposal to launch *Triana*—a satellite named after Rodrigo de Triana, the lookout on Columbus's ship who first

sighted the New World. See Kathy Sawyer, "The World, Live—Just a Click Away," *Washington Post*, March 13, 1998, p. A1. *Triana*, a kind of "electronic Magellan" for the twenty-first century, would send round-the-clock live coverage of the planet to a ground station network called Earth-Space, which would transmit images to anyone with television or Internet access. Gore's March 1998 announcement prompted NASA to try and make the "all-Earth, all-the-time" images available by the year 2000. When Gore's proposal was discussed during a congressional hearing in March 1998, a member of the Federation of American Scientists scoffed and called it nothing more than "performance art," claiming it lacked scientific justification. See Elizabeth Shogren, "Gore Prods NASA for Live Coverage of Earth," *Los Angeles Times*, March 14, 1998, p. A22. As the mockery of Gore's proposal suggests, the prospect of a public remote-sensing satellite is at the least distasteful to military and scientific elites, who have historically had a monopoly on the knowledges generated from an orbiting vantage point. Although Gore's proposal is without a doubt expensive (and perhaps a waste of the nation's financial resources, especially given its duplication of other similar satellites), it is the first official proposal for a "public access" remote-sensing satellite, for *Triana* was designed to make satellite images (however generic) widely and instantly available to the public free of charge.

10 Anselmo, "Remote Sensing to Alter TV News," p. 61.

11 Ibid.

12 Ibid.

13 See, e.g., *Safe Haven: The United Nations and the Betrayal of Srebrenica*, video dir. Ilan Ziv, 1996; *Srebrenica: A Cry from the Grave*, film dir. Leslie Woodhead, 1997; Robert Block, "The Fall of Srebrenica," *The Independent London*, Oct. 30, 1995; Honig and Both, *Srebrenica*; Rohde, *Endgame*; Sudetic, *Blood and Vengeance*; Stover and Peress, *The Graves*; and McCarthy, *After the Fall*.

14 For a discussion of the way the Serbs have been positioned in Western cultural discourse see chap. 9, "Demonizing the Serbs," in Parenti, *To Kill a Nation*; and Longinovic, "Vampires like Us."

15 Corwin, *Dubious Mandate*; also see the online report "Ethnic Conflicts in Civil War in Bosnia."

16 "War Crimes Prosecutor Wants Team to Hunt Suspects"; Graff, "Commanders in Court."

17 Associated Press, "Rights Groups Demand Release of Srebrenica Data," April 4, 1996; and Rohde, *Endgame*.

18 Ignatieff, *Virtual War*.

19 Fiske, *Media Matters*, pp. 7–8.

20 Zumach, "US Intelligence Knew Serbs Were Planning an Assault on Srebrenica."

21 See Corwin's *Dubious Mandate* for information about the Muslim government's manipulations of Srebrenica's civilians.

22 Stephen Engelberg and Tim Weiner, "Days of Slaughter: The Killing of Srebrenica," *New York Times*, Oct. 29, 1995, p. 1.

23 United Nations, Office of the Secretary-General, "Report of the Secretary-General on Bosnia and Herzegovina."

24 David Rohde, "Eyewitnesses Confirm Massacres in Bosnia," *Christian Science Monitor*, Oct. 5, 1995, p. 1.

25 Stephen Engelberg and Tom Weiner, "Days of Slaughter: The Killing of Srebrenica," *New York Times*, Oct. 29, 1995, p. 1.

26 For a detailed discussion of the contradictory stories and the strategic use of numbers see Ryan, "What's in a Mass Grave?"

27 Israel and Malic, "Falsely Accused." Also see Antic, "Srebrenica Massacre Denial."

28 For a thorough discussion of the politics of ethnic cleansing see Woodward, "War: Building States from Nations." Complicating Western assumptions on the matter, Woodward argues that "the expulsion of persons according to ethnic background, which came to be labeled ethnic cleansing, had nothing to do with ethnicity, but rather with securing national rights to land. And because the resulting war is waged to define who can belong to a particular state and its territory, it makes no distinction between soldiers and civilians, between military and civilian targets" (p. 216). She continues: "The basis of this policy of ethnic cleansing lay not with primordial hatreds or local jealousies, but with political goals" (p. 222).

29 The Red Cross has published two books featuring photographs of human remains found in mass-grave sites near Srebrenica, hoping that family members could use them to identify the dead. Associated Press, "Bodies from Srebrenica Massacre Go Unclaimed," *Los Angeles Times*, May 15, 2001, p. A4.

30 Sells, "Seeking the Truth of Srebrenica."

31 For more detailed accounts of the Srebrenica massacre see Rohde, *Endgame*.

32 Gow, Paterson, and Preston, introduction to *Bosnia by Television*, p. 3.

33 For further discussion of the history and critique of tele-presence and tele-action see Goldberg, *Robot in the Garden*; and Manovich, *The Language of New Media*.

34 Raymond Williams offers the term "flow" to describe commercial television's structure (Williams, *Television*, p. 74), but he does not explore the particularities of segments that constitute that flow. The term coverage is a useful supplement to "flow" because it implies the need to consider the various perspectives or views that are integrated within and in turn work to shape the medium's content and form. Television studies needs to cultivate more sensitivity to what Dick Hebidge calls "the incandescence of the particular" (Hebidge, *Hiding in the Light*, 1988, p. 230), especially since television is so often reduced, whether to a medium of flow, a technology of commercial entertainment, or a form of culture that lacks subtlety, complexity, and potential.

35 Live international television coverage emerges during the medium's first decade with such series as NBC's *Wide Wide World* (1954–1958). For further discussion see Parks, "As the Earth Spins." The filmic predecessor to global television "coverage" is, of course, the film travelogue of the early twentieth century. See Griffiths, " 'To the World the World We Show,' " pp. 282–307.

36 Weber, *Mass Mediauras*, p. 122.

37 The *New York Times*, the *Christian Science Monitor*, and Amnesty International published comprehensive stories on the Srebrenica massacres and reprinted

satellite images released by the U.S. State Department. These images also appeared on an Internet site maintained by the Federation of American Scientists. See "The Bosnia Massacre."

38 One might consider the integration of this satellite intelligence on television in relation to the use of satellite weather photos, which must always be "anchored" and interpreted by a meteorologist. For a discussion of the discourse of weather forecasting see Ross, *Strange Weather*.

39 Weber, *Mass Mediauras*, p. 122.

40 *Tromp l'oeil* is a French phrase used to connote a surprising and unexpected "blow to the eye" or the act of fooling or tricking the eye.

41 Horace Newcomb defines television as a medium of intimacy. See Newcomb, "Toward a Television Aesthetic."

42 For a discussion of remote-sensing images, weather, and nationalism see Berland, "Mapping Space." For discussion of the way that "civilizational" discourses have been applied to the former Yugoslavia to ostracize and demonize the Serbs see Longinovic, "Music Wars."

43 Hozic, "Making of the Unwanted Colonies," p. 237. For a discussion of the coverage of Rwandan refugee critics see Fair and Parks, "Africa on Camera."

44 William J. Mitchell discusses the changing status of truth in the context of digitization in his book *The Reconfigured Eye*.

45 Feuer, "The Concept of Live Television"; Mellencamp, *High Anxiety*.

46 The sheer volume of satellite image data gathered since the 1960s prompted President Clinton to commercialize the remote-sensing industry in 1994. Some speculated this would lead to an increase in the mass circulation of satellite images on television. See, e.g., Anselmo, "Remote Sensing to Alter TV News."

47 Weber, *Mass Mediauras*, pp. 122–123.

48 Rapaport, "World War 3.1," p. 126. For a broader evaluation of information warfare see Buchan, "Information War and the Air Force."

49 Barella, "Givin' 'em Their Space."

50 Rapaport, "World War 3.1," p. 129.

51 These "sector searches" have also been commissioned by the International War Crimes Tribunal. See Capaccio and Greczyn, "Warfare in the Information Age."

52 Rapaport, "World War 3.1," p. 125.

53 Richelson and Gordon, "High Flyin' Spies."

54 Rapaport, "World War 3.1," p. 127.

55 Barella, "Givin' 'em Their Space."

56 Ibid.

57 Ignatieff, *Virtual War*, p. 196.

58 Numerous American TV news broadcasts after the Srebrenica massacre positioned American political officials and troops as part of a cleanup process and ongoing investigation into the atrocities. Reports on CNN and ABC showed people using "earth movers," "metal detectors," and "chopsticks" to pick through debris during the excavation of mass-grave sites.

59 See Sylvia Poggioli, "Scouts without Compasses," Nieman Foundation, Harvard

University, *Nieman Reports* 53–54 (1999–2000). Also see Gow et al., *Bosnia by*
Television.

60   Virilio, *Strategy of Deception*.

61   Kellner, *The Persian Gulf TV War*.

62   For a discussion of the historical relations between war and imaging technologies see Virilio, *War and Cinema*.

63   Robins, *Into the Image*, p. 64.

64   Ibid., pp. 63–64.

65   Ignatieff, *Virtual War*, p. 196.

66   Ibid., p. 4.

67   Virilio uses the phrase "eyeless vision" in *War and Cinema*, suggesting, "Alongside the 'war machine,' there has always existed an ocular . . . 'watching machine'
capable of providing soldiers, and particularly commanders, with a visual perspective on the military action under way. From the original watch tower through
the anchored balloon to the reconnaissance aircraft and remote-sensing satellites, one and the same function has been indefinitely repeated, the eye's function being the function of a weapon" (*War and Cinema*, p. 3). Eyeless vision thus
refers to the automated gaze of military machines such as orbiting satellites,
which not only watch but, in doing so, also function as weapons.

68   Mclear, *Beclouded Visions*, p. 5.

69   Ibid., p. 4.

70   Ibid., p. 16.

71   Ibid., p. 14.

72   Ibid., p. 12.

73   David Rohde, "How a Serb Massacre was Exposed: *Monitor* Reporter Eluded
Soldiers and Discovered Evidence of Serb Atrocities," *Christian Science Monitor*,
Aug. 25, 1995, p. 1.

74   David Rohde, "Eyewitness Report Supports Charges by US of Killings," *Christian Science Monitor*, Aug. 18, 1995, p. 1. Rohde won a 1996 Pulitzer Prize for his
coverage of the Srebrenica massacres and has since published the book *Endgame:
The Betrayal and Fall of Srebrenica, Europe's Worst Massacre since World War II*.

75   Ibid.

76   Rohde, "How a Serb Massacre Was Exposed: *Monitor* Reporter Eluded Soldiers
and Discovered Evidence of Serb Atrocities," *Christian Science Monitor*, Aug. 25,
1995, p. 1; and Peter Grier, "Monitor Correspondent Wins Pulitzer," *Christian
Science Monitor*, April 10, 1996, p. 1.

77   Gonzalez, "Autotopographies," pp. 133–134.

78   Ibid., p. 134.

79   Barthes, *Mythologies*.

80   Silverman, *The Threshold of the Visible World*, p. 156.

81   See, e.g., the work of the Public Eye initiative, which purchases satellite images
to expose state secrets and strategic silences: http://www.fast.org/eye.

82   Weber, *Mass Mediauras*, pp. 118–119.

83   As John Fiske suggests, the media event "also invites intervention and motivates

people to struggle to redirect at least some of the currents flowing through it to serve their own interests; it is therefore a site of popular engagement and involvement, not just a scenic view to be photographed and left behind. Its period of maximum visibility is limited, often to a few days, though the discursive struggles it occasions will typically continue for much longer" (John Fiske, *Media Matters*, pp. 7–8).

84 Silverman, *The Threshold of the Visible World*, pp. 160–161.

85 Ibid.

86 "Message from the Women of Srebrenica in Favor of a Memorial in Potocari"; also see "Women of Srebrenica."

87 "Bosnian Muslim Refugees Visit Homes in Srebrenica Area." This was also confirmed in my conversations with UN workers who are now living in Srebrenica and helping facilitate the return of Muslim residents.

88 See Sells, "Seeking the Truth in Srebrenica."

89 Vismann, "Starting from Scratch," pp. 47, 46.

Chapter 4. Satellite Archaeology: Remote Sensing Cleopatra in Egypt

1 Foucault writes, "To substitute for the enigmatic treasure of 'things' anterior to discourse, the regular formation of objects that emerge only in discourse. To define these objects without reference to the ground, the foundation of things, but by relating them to the body of rules that enable them to form as objects of a discourse and thus constitute the conditions of their historical appearance. To write a history of discursive objects that does not plunge them into the common depth of a primal soil, but deploys the nexus of regularities that govern their dispersion" (Foucault, *The Archaeology of Knowledge*, pp. 47–48).

2 Holland, "Cleopatra: What Kind of Woman Was She Anyway?" p. 56.

3 Hughes-Hallet, *Cleopatra*.

4 In Alexandria remote sensing has been used to reaffirm origin myths of Western civilization, even as historiography of the classical period has been challenged by feminist and postcolonialist critiques. As postcolonial critic Dympna Callaghan reminds us, "We have to be aware of *how* we get our information about her [Cleopatra] and why it is structured quite in the way it is" (Callaghan, "Representing Cleopatra in the Postcolonial Moment," p. 41).

5 Joyce, "Archaeology Takes to the Skies," p. 42.

6 Wellborn, "History's Secrets Yielding to High-Tech Devices," p. 68.

7 The first known aerial photographs of an archaeological site were pictures of Stonehenge taken from a war balloon by Lieutenant P. H. Sharpe in the early 1900s. See "The Sky's Eyes."

8 Ibid.

9 Joyce, "Archaeology Takes to the Skies." See also Huseonica, "Archaeology from Above." A decade before Sever's work, the *Salt Lake City Tribune* published an article with the headline "Noah's Ark 'Sighted' by Satellite" (Feb. 21, 1974), which discussed the Earth Resource Technology Satellites' discovery of a mass that ap-

peared to be Noah's ark on a military reserve near the Russian and Iranian borders. "Satellites and Archaeology" file, NASA History Office.

10   Lightfoot and Lightfoot, "Revealing the Ancient World through High Technology," p. 54.

11   John Noble Wilford, "Lofty Instruments Discern Trace of Ancient Peoples," *New York Times*, March 10, 1992, C1.

12   Tom Sever, interview by Neil McAleer, p. 71.

13   Joyce, "Archaeology Takes to the Skies," p. 42.

14   Allman, "Finding Lost Worlds," p. 69.

15   Ibid.

16   Gibbons, "A 'New Look' for Archaeology."

17   Ibid. Also see Lightfoot and Lightfoot, "Revealing the Ancient World."

18   Schaefer, "NASA's Earth Observation System Data Information System."

19   Joyce, "Archaeology Takes to the Skies," p. 42.

20   Sever, interview by McAleer, p. 71.

21   Wellborn, "History's Secrets Yielding to High-Tech Devices," p. 68.

22   Many of these excavations were featured in a news clipping I found in the "Satellite and Archaeology" file at the NASA History Office. NASA TV has featured video of satellite images of some of these sites (e.g., the Great Wall in China in April 1996; Angkor, Cambodia; the "Lost City of Ubar" in Oman; and the Silk Road in the northwest China desert).

23   Sever, interview by McAleer, p. 71.

24   Berland, "Mapping Space," p. 127.

25   Hartley, *The Uses of Television*, p. 18.

26   Wellborn, "History's Secrets Yielding to High-Tech Devices," p. 68.

27   Gibbons, "A 'New Look' for Archaeology," p. 918.

28   Joyce, "Archaeology Takes to the Skies," p. 42.

29   Wellborn, "History's Secrets Yielding to High-Tech Devices," p. 68.

30   Lightfoot and Lightfoot, "Revealing the Ancient World through High Technology."

31   Keizer, "Stones on the Screen," p. 8.

32   Lightfoot and Lightfoot, "Revealing the Ancient World through High Technology," p. 54.

33   For a critique of the God's-eye view of science see Haraway, *Simians, Cyborgs, and Women*, pp. 183–201.

34   In some cases television is also becoming more archaeological. Fictional shows like *Xena: Princess Warrior*, for instance, can be seen as "excavating" the ancient world without the pretense of science. One episode of the show features Xena avenging the death of Cleopatra by killing Marc Antony to restore Egypt's political independence. I thank Kevin Glynn for sharing this insight with me.

35   Foucault, *The Archaeology of Knowledge*, p. 195.

36   Within classical Hollywood cinema Cleopatra has certainly been positioned as a sexual spectacle. For a discussion of sexual representation and the cinematic apparatus see Laura Mulvey's "Visual Pleasure and Narrative Cinema," pp. 28–40.

37   Holland, "Cleopatra," p. 56.

38   Review of *Cleopatra*, p. 122.

39   Ibid.

40   Hamer, *Signs of Cleopatra,"* p. 118.

41   Higashi, "Antimodernism as Historical Representation in a Consumer Culture,"
     p. 91.

42   Higham, *Cecil B. DeMille*, p. 231.

43   Ibid.

44   Ibid.

45   Review of *Cleopatra*.

46   Shohat and Stam, *Unthinking Eurocentrism*, pp. 144–145.

47   Gundle, "Sophia Loren, Italian Icon," p. 371.

48   In 1989 Italian officials proclaimed Sophia Loren "a timeless symbol of her coun-
     try's spirit" and proposed erecting a statue in her honor. Gundle, "Sophia Loren,
     Italian Icon," p. 367. *Due Notti* also alluded to Cleopatra's racial ambiguity by
     creating blonde and brunette incarnations of the queen.

49   Sophia Loren was also considered for this Cleopatra role as was a litany of other
     actresses, including Marilyn Monroe, Brigitte Bardot, Rita Hayworth, Gina Lollo-
     brigida, Jennifer Jones, and Audrey Hepburn. See Tony Crawley, *The Films of
     Sophia Loren*, p. 60.

50   Quoted in Jack Brodsky and Nathan Weiss, *The Cleopatra Papers*, p. 109.

51   Ibid.

52   Ibid.

53   Lefkowitz, "Not Out of Africa," p. 29. See also Begley, "Out of Egypt, Greece,"
     p. 49; and Elson, "Attacking Afrocentrism," p. 66.

54   Burton and Taylor had an impassioned affair on the set and attracted paparazzi,
     who snuck onto the set in Rome and snapped pictures of Taylor. The scandals
     surrounding the production of *Cleopatra* only heightened the film's spectacular
     appeal.

55   Hamer, *Signs of Cleopatra*, p. xvii.

56   The exhibit traveled from Paris to Ottawa to Vienna. See Porterfield, "Egyptoma-
     nia!" p. 84.

57   Hamer, *Signs of Cleopatra*, p. xix.

58   Cited in May, "The Circuitous Route of Edmonia Lewis' Masterwork," p. 16.

59   Ibid.

60   Hunter-Gault, "Testament to Bravery."

61   Modern historiography on ancient civilization, Bernal explains, was shaped by
     the scientific racism of nineteenth-century intellectual movements. He writes:
     "With the rise of a passionate and systematic racism in the 19th century, the
     ancient notion that Greece was a mixed culture that had been civilized by Afri-
     cans and Semites became not only abominable but unscientific" (Bernal, *Black
     Athena*, p. 441).

62   Ibid., p. 2.

63   Ibid.

64   Lefkowitz, *Not Out of Africa*, pp. xv–xvi.

65   The September 23, 1991, cover of *Newsweek* featured an image of the Egyptian queen from a temple relief wearing a modern earring superimposed by the question: "Was Cleopatra Black?" Some Afrocentrists have suggested that since so little is known about Cleopatra's paternal grandmother (except that she was not legally married to Ptolemy IX), it is possible that she was of African descent.

66   Lefkowitz, *Not Out of Africa*, p. 51.

67   See, e.g., a discussion of blaxploitation and *Cleopatra Jones* in Morris, "Blaxploitation."

68   I thank Jennifer Fuller for the information about Raven Symone, who apparently announced the name of her dog when she was a guest on the *Arsenio Hall Show*.

69   Phyllis Rose, *Jazz Cleopatra: Josephine Baker in Her Time* (New York: Doubleday, 1989).

70   Holt, "Why We Need a Black Press."

71   Hamer, *Signs of Cleopatra*, p. xviii.

72   McClintock, *Imperial Leather*, pp. 22, 23. I thank Ursula Biemann for encouraging me to make this connection.

73   Alexandria was founded by Alexander the Great in 332 BC. The city was a vital trading port famed for its massive library, the Pharos, and the lighthouse, a four-hundred-foot tower considered one of the Seven Wonders of the World. These underwater ruins form one-third of Ptolemaic Alexandria. The city was supposedly destroyed by a tidal wave following an earthquake in AD 355.

74   Scheherezade Faramarzi, "Undersea Archaeologists Map Court of Cleopatra," *Washington Times*, Nov. 4, 1996, p. A1.

75   Michael Murphy, "Cleopatra's Playground Revealed," *The Times of London*, Nov. 4, 1996.

76   "Sunken Treasures," p. 6.

77   Murphy, "The Royal City Beneath the Sea," *The Times of London*, Nov. 9, 1996.

78   "Marine Archaeologist Reveals Cleopatra's Preserved Palace," *Deutsche Presse-Agentur*, Nov. 3, 1996.

79   Murphy, "The Royal City Beneath the Sea"; and Robert Uhlig, "City Beneath the Sea Surrenders Its Secrets: Satellite Technology Helps Solve Enigma of Alexandria," *Daily Telegraph*, London, Nov. 4, 1996, p. 9.

80   Scheherezade Faramarzi, "Undersea Archaeologists Map Court of Cleopatra," *Washington Times*, Nov. 4, 1996, p. A1.

81   Michael Murphy, "Cleopatra's Playground Revealed," *The Times of London*, Nov. 4, 1996.

82   Ibid. See also "Divers Discover Cleopatra's Alexandrian Trysting Place," *The Guardian*, Nov. 4, 1996, p. 11; and Robert Uhlig, "City Beneath the Sea Surrenders Its Secrets," *Daily Telegraph*, London, Nov. 4, 1996, p. 9.

83   Peggy Brown, "Cleopatra's Alexandria Revealed the Romance of a Lost City," *Newsday*, Nov. 12, 1996, p. A31.

84   "Marine Archaeologist Reveals Cleopatra's Preserved Palace," *Deutsche Presse-Agentur*, Nov. 3, 1996.

85   Ibid.

86   "Sunken Treasures," p. 6.

87 Young, "Divers Recover Ancient Wonders."

88 *Treasures of the Sunken City* is a NOVA documentary that first aired on PBS on Nov. 18, 1997.

89 Murphy, "The Royal City Beneath the Sea."

90 "Mapping the Treasures."

91 Many speculate that this satellite excavation will lead to new revisionist accounts of the last pharaonic phase. A professor of classical archaeology at Alexandria University, Dr. El Faharani, claims, "Until we had seen this map, ancient Alexandria was an enigma. Previous maps were a guess. This map is accurate to the inch, showing quays and numerous artifacts, and indicating where streets could have run and palaces stood" (quoted in Robert Uhlig, "City Beneath the Sea Surrenders Its Secrets," *Daily Telegraph*, London, Nov. 4, 1996, p. 9). And the European Institute of Marine Archaeology released a statement claiming, "The exact topography of the vanished royal city can be identified for the first time. From now on, the accurate maps resulting from this mission will form the basis for all future archaeological work in this area" ("Divers Discover Cleopatra's Alexandrian Trysting Place," *The Guardian*, Nov. 4, 1996, p. 11). One writer claimed, "The new palace map will force extensive rewriting of archaeological theories" (James Cusick, "Sea Gives Up Lost Secrets of Alexandria," *The Independent*, London, Nov. 4, 1996, p. 10). And Professor Aziza said, "This discovery is quite revolutionary. It will allow us to rewrite in detail the last days of the pharaohs" ("Divers May Have Found Lighthouse," *Rocky Mountain News*, Nov. 3, 1996, p. 54A). Although this evidence will likely lead to a revision of conventional ancient history, it is unclear just how it may relate to Afrocentrist revisionism. For by asking different questions and beginning with different assumptions about Western progress, Egyptology in general may be ill equipped to address the issues of cultural difference posed by Afrocentrism.

92 Marvin, *When Old Technologies Were New*, p. 109.

93 Bryld and Lykke, *Cosmodolphins*, p. 21.

94 Foucault, *The Archaeology of Knowledge*, p. 48.

95 Mitchell, *The Reconfigured Eye*, p. 57. For other work on the senses and media see Marks, *Touch* and Morse, *Virtualities*.

Chapter 5. Satellite Panoramas:
Astronomical Observation and Remote Control

1 Chaisson, *Hubble Wars*.

2 For a description of the instruments on board the Hubble Space Telescope see "General Overview of the Hubble Space Telescope."

3 Kathy Sawyer, "Given New Focus, Hubble Can Almost See Forever," *Washington Post*, Jan. 14, 1994, p. A1.

4 Ibid.

5 Terence Dickinson, "Fixed Hubble Could Rewrite the Textbooks," *Toronto Star*, Jan. 30, 1994, p. E9.

6 The Space Telescope Science Institute (STSCI) conducts and coordinates the sci-

entific operations of the Hubble at Johns Hopkins University and is overseen by the Association of Universities for Research in Astronomy, Inc. (AURA). Data can be broadcast from Hubble to the ground stations immediately or stored on tape and downlinked later. The observer on the ground can examine the "raw" images and other data within a few minutes for a quick-look analysis. Within twenty-four hours GSFC formats the data for delivery to the STSCI, where it is processed and maintained for the scientific community.

7   See Spigel, "From Domestic Space to Outer Space," p. 228.

8   Dean, *Aliens in America*, p. 68.

9   Levy, *Comets*, p. 208.

10 Ibid.

11 "Planetary Punishment: Comet to Spank Jupiter's Behind," *Salt Lake Tribune*, July 16, 1994.

12 *CNN Headline News*, July 13, 1994.

13 "Comet Hitting the Earth Would Equal 15 Million Hiroshima Bombs," *Agence France Presse*, April 18, 1997.

14 Gwynne Dyer, "Somewhere, Tonight, a World Is Ending," *Baltimore Sun*, July 13, 1994, p. 13A.

15 Slim Randles, "New Mexico's a Great Place to Watch the Skies," *Albuquerque Journal*, Feb. 12, 1998, p. 4.

16 The company also initiated a promotion that made the client the owner of the Shoemaker-Levy comet, calling it the "crash of 94." Rich, "High Profile Clients; Low Profile Employment," p. 24.

17 *CNN Headline News*, July 13, 1994.

18 Buck-Morss, "Art in the Age of Its Electronic Production," p. 6.

19 Two years after Shoemaker-Levy crashed into Jupiter, the media industries began capitalizing on popular interest in the apocalyptic possibility that an asteroid or comet might plunge to Earth. NBC estimates that seventy million people watched all or part of *Asteroid*, a two-part, four-hour "sweeps disasterfest." It was the highest rated TV movie of the season and even beat out the television premiere of Disney's *The Lion King*. Television documentaries on asteroids appeared in early 1997 as well. NBC broadcast a *National Geographic* special called *Asteroids: Deadly Impact*, and the Discovery Channel showed a two-hour documentary called *Three Minutes to Impact*. See "Real Disasters Reviewed," *Arkansas Democrat-Gazette*, Feb. 25, 1997, p. 3E. This fascination with terrestrial collisions culminated in Hollywood's 1998 summer releases *Deep Impact* and *Armageddon*, which feature comets or asteroids destroying parts of Earth. And four years after Jupiter's collision with Shoemaker-Levy 9, Hubble images of planetary crashes were supplemented by even more grandiose *galactic* collisions. As one journalist described it, Hubble "has peered farther than ever before into the heart of a giant galaxy that is smashing into—and eating—a smaller galaxy, to reveal the fiery maelstrom around a monstrous black hole that is feeding off the cosmic carnage" (Kathy Sawyer, "Hubble Spies Black Hole Feeding Off Galaxy," *Washington Post*, May 15, 1998, p. A3).

20 Mellencamp, *High Anxiety*, p. 150.

21   Ibid.

22   Parks, "As the Earth Spins."

23   See Mitchell's *The Reconfigured Eye*. Mitchell argues that digitization destabilizes the truth status of images and encourages us to be more "knowing" about the status of images in general. So too, I would argue, do technological convergences. In the case of Hubble the remoteness of the telescope's vision and the precariousness of its transmissions may raise as much uncertainty about the image's truth status as digitization does. Thus while digitization may prompt us to be more "knowing" about the status of images, this tendency is especially pronounced in relation to Hubble's renderings of otherworldly phenomena since they are not just digital images but telescopic transmissions. In other words, it's important to consider how specific technological convergences alter the ontological status of the image, which may or may not hinge on the digital.

24   Feuer, "The Concept of Live Television," p. 15.

25   "Hubble's Deepest View of the Universe Unveils Bewildering Galaxies across Billions of Years."

26   Quoted in Cowen, "After Hubble: The Next Generation," *Science News*, April 26, 1997, p. 262.

27   Ibid., p. 95.

28   Ibid., p. 100.

29   While Hubble images have been used to "find" human origins in deep space, they have also exposed stellar conflicts and disruptions. *Science News*, for instance, published an article that oddly transposes anxieties about human abortion onto the process of star formation in deep space. Describing activity in the Eagle and Orion nebulae featured with the article, Ron Cowen writes, "Somewhere in the spiral reaches of our galaxy, a stellar womb slowly assembles. As this cloud of dust and cold hydrogen gas draws together an embryo takes shape deep inside. A dense ball surrounded by a vast disk, the embryo grows hotter and more compact as gravity pulls more and more material from its mother cloud. Suddenly, the womb disintegrates. Alone in the harsh environment of interstellar space, its supply line severed, the fledgling star can grow no bigger. Although it may go on to shine for several billion years, the star's premature birth has irrevocably altered its destiny" (Cowen, "Bullies of the Universe," p. 350).

30   Foucault, *The Archaeology of Knowledge*, p. 205.

31   Adler, "Witness at the Creation," p. 71.

32   Naeye, *Through the Eyes of Hubble*, pp. 40–42.

33   Lemonick, "Cosmic Close-Ups," p. 95.

34   Ibid.

35   Reston, "Orion: Where Stars are Born," pp. 91–92. *National Geographic*, renowned for its appropriation of remote territories and peoples, claims deep space matter as a domain of Western vision, knowledge, and territory. For a discussion of such practices see Lutz and Collins, *Reading "National Geographic."* I thank Sujata Moorti for encouraging me to make this connection.

36   Begley, "Postcards from the Edge," p. 54.

37   Cowen, "After Hubble," p. 262.

38   Ibid.

39   Begley, "When Galaxies Collide," p. 33.

40   The Space Telescope Science Institute celebrates Hubble's ability to reveal "gal-
       axies under construction in the early universe, out of a long sought ancient
       population of 'galactic building blocks' " ("Hubble Sees Early Building Blocks of
       Today's Galaxies"). Astronomers treat ancient stars as ruins or artifacts: "Like a
       well-preserved fossil, an ancient star whose surface remains unchanged should
       reflect the chemical composition characteristic of the universe soon after the
       Big Bang" (Cowen, "Hubble," p. 79). Astronomer Rodger Thompson insists that
       Hubble's NICMOS (Near-Infrared Camera, installed in 1997) "allows us to see a
       'fossil record' of the star's late evolutionary stages" ("Hubble's Upgrades Show
       Birth and Death of Stars; Discover Massive Black Hole").

41   As Carol Stabile explains, visual technologies like the sonogram "isolate the em-
       bryo as astronaut, extraterrestrial, or aquatic entity" and keep the mother's body
       out of view, erasing "the female body's contribution in terms of labor to human
       reproduction." See Stabile, "Shooting the Mother," p. 189.

42   Penley, "Time Travel, Primal Scene and the Critical Dystopia," p. 120.

43   "Hubble's Deepest View of the Universe Unveils Bewildering Galaxies across
       Billions of Years."

44   Kember, "Feminist Figuration and the Question of Origin," p. 264.

45   "*Cosmic Voyage*: Production Firsts and Facts."

46   "The Making of *Cosmic Voyage*."

47   "Scientific Visualization in the IMAX Film *Cosmic Voyage*."

48   "Scientific Visualization Reaches New Heights in IMAX Film."

49   Since this is a digital effect, the camera, of course, does not exist, but its position,
       or perspective, is implied.

50   Walter Benjamin, *Illuminations*, pp. 259–260.

51   We hear snippets of audio from the song "Funkytown"; the themes to *Dallas*,
       The Andy Griffith Show, and *The Twilight Zone*; the pop tunes "Itsy Bitsy Teeny
       Weeny Yellow Polka Dot Bikini," "Volare," and "Catch a Falling Star"; an Al-
       mond Joy commercial; President Nixon's press conferences; a Martin Luther
       King Jr. speech; the John F. Kennedy assassination; and the HUAC hearings,
       among others.

52   Some religious cults posit mythologies about the human souls of the deceased
       spending time in "special interplanetary stations, being stripped of memories
       and deprived of power, before being inserted into a new body perhaps on an alien
       planet" (Bainbridge, "Religions for a Galactic Civilization," p. 197).

53   Ellie's personal narrative of space travel reproduces some of the tenets of Scien-
       tology, a religious discourse that places humans in a galactic context. William
       Sims Bainbridge explains: "As in psychoanalysis, one of the main techniques of
       Scientology has people enter the world of memory, recalling early traumatic inci-
       dents which supposedly left subconscious memory traces called engrams. These
       engrams reduce the person's level of effectiveness in the present, and must be
       removed" (Bainbridge, "Religions for a Galactic Civilization," p. 197).

54   We might consider the alien forms designed by digital artist Marcos Novak as

useful analogues for spurring ways of engaging with Hubble images. In a work called "TransAura" Novak creates a series of alien illusions. He uses algorithms to generate alien forms, builds physical sculptures of those forms, digitally photographs them, and superimposes massive versions of them onto images of public settings, constructing the appearance that they were actual physical sculptures in time and space. Novak describes this as a technique of "eversion" since he uses the virtual not as an "immersive" space but rather to derive an element that can become physical and then trigger the viewer to imagine something that cannot be seen. The work establishes a set of conceptual relations between the virtual, the physical, and the invisible, moving matter between these fields of (un)intelligibility. Novak's alien forms stretch our perceptual capacities by prompting us to imagine quintessentially digital forms of a massive scale as part of earthly environments. This, it seems to me, is precisely the kind of imaginary practice that Hubble images can open up and ask of us but that has been repeatedly infused with linear visualizations that reinforce evolutionary humanism. For further discussion and project prototypes see Marcos Novak's Web site: http://www.centrifuge.org/marcos/ (accessed Oct. 10, 2002).

55   As Foucault explains, the role of an archaeology of knowledge "is not to dissipate oblivion, to rediscover, in the depths of things said, at the very place in which they are silent, the moment of their birth (whether this is seen as their empirical creation, or the transcendental act that gives them origin); it does not set out to be a recollection of the original or a memory of the truth. On the contrary, its task is to *make* differences: to constitute them as objects, to analyse them, and to define their concept" (Foucault, *The Archaeology of Knowledge*, p. 205).

56   The Search for Extra Terrestrial Intelligence project or SETI emerged in 1960 when a group of radio astronomers interested in "interstellar communication" began to seek evidence of intelligent life elsewhere in the universe. For more than forty years SETI scientists have used radio astronomy technologies in an effort to locate extraterrestrial intelligence. In 1993 Congress terminated all federal funding for SETI projects administered through NASA. The SETI projects have continued, however, and have been funded through private donations. Arthur Clarke sat on SETI's advisory board, and Carl Sagan was directly involved in the formation of the SETI Institute. Also see, "What Is SETI?" SETI worked with Warner Brothers on the production of the film *Contact*, which is based on Sagan's *Contact: A Novel*. For a critical discussion of Carl Sagan's relationship to popular culture see Penley, *NASA/TREK*, pp. 5–10. Also, for a fascinating discussion of SETI see Peters, *Speaking into the Air*, pp. 246–257.

57   Federally owned and administered by the National Science Foundation, the VLA collects radio signals emitted by astronomical bodies in frequencies between 74 and 50,000 MHZ. Its construction was approved by Congress in 1972, funded by U.S. taxpayers, and finished in 1980. Since then, astronomers from around the world have used it for atmospheric studies, satellite tracking, and sky mapping.

58   For discussion of weather and global politics see Ross, *Strange Weather*, pp. 204–213.

59   For discussion of the Global Warming Prevention Conference see "Kyoto Proto-

col on Climate Change Opens for Signature." Many developing nations complained that they should not have to reduce emissions rates as much as the United States because they do not contribute as large a share to the pollution. See, e.g., Masood, "Equity Is the Key for Developing Nations."

60 Guillermo Gomez-Peña suggests that immigrants are positioned in the United States as "invaders from the South, the human incarnation of the Mexican fly, subhuman 'wetbacks,' the 'alien' from another (cultural) planet. They are accused of stealing 'our jobs,' of shrinking 'our budget,' of taking advantage of the welfare system, of not paying taxes and of bringing disease, drugs, street violence, foreign thoughts, pagan rites, primitive customs and alien sounds" (Gomez-Peña, *New World Border*, p. 67).

61 The transmission of electronic signals at the borders—whether high-powered "border blaster" radio stations in the 1950s or transnational satellite television services like Telemundo or Globovision in the 1990s—provokes nationalist anxieties about signal spillover and cultural invasion. For a history see Fowler and Crawford, *Border Blasters*.

62 For a discussion of border militarization projects in Arizona, Texas, and California around the time of this film's release see Parenti, "Crossing Borders," p. 15.

63 For further discussion of the remote control see Bellamy and Walker. *Grazing on a Vast Wasteland*; Ferguson, "Channel Repertoire in the Presence of Remote Control Devices, VCRS, and Cable Television"; and Walker and Bellamy, *The Remote Control in the New Age of Television*.

64 Seiter et al., introduction to *Remote Control*, p. 2.

65 Newcott, "Time Exposures," p. 3.

66 "Who Art in Heaven?" p. 23.

67 As Kevin Glynn suggests, such "tabloid knowledges often express an ambivalent and conflicted incredulity toward scientific rationalism and its power to produce authoritative truths" (Glynn, *Tabloid Culture*, p. 96).

68 As Richard Panek reminds us in *Seeing and Believing*, "The technology of the telescope had always set limits on the available information, and observers had always used those limits to reinforce prevailing beliefs even while advancing new ones" (p. 174).

69 John Fiske explains, "Disrupting the centeredness of the grand narrative is a useful deconstructive move, but it is not, in itself, enough: it may dislodge some of the certainties of the current regime of power, but it does little to promote the next unless some further steps are taken" (Fiske, *Power Plays, Power Works*, p. 289).

Conclusion

1 For further discussion of Nam Jun Paik's work see Hanhardt et al. *The Worlds of Nam Jun Paik*.

2 Cited in MacKenzie and Wajcman, *The Social Shaping of Technology*, p. 10.

3 Anthony Giddens argues that the potential for totalitarianism is a permanent feature of late capitalism. He suggests that the term *totalitarian* should be used

to describe tendencies toward a particular form of state rule. These tendencies include an emphasis on surveillance, moral totalism, terror (i.e., maximal police power combined with industrialized war-waging and sequestration) and the prominence of a single leader who generates mass support. See Giddens, *The Nation-State and Violence*, pp. 303–304.

4   For more information about Deep Dish TV see Deep Dish TV home page.

5   See ECI Telecollaborative Art Projects.

6   Ibid. In this satellite project performance is articulated as a kind of "out-of-body experience" in which the body takes on the form of an electronic signal circulating across the planet, constantly coming into being by virtue of its own relation to the processes of signal transmission, circulation, and scanning.

7   Marko Peljhan, interview by Rasa Smite and Ieva Auzina, World-Information .Org Web site.

8   Peters, *Speaking into the Air*, p. 180. For a discussion of the "quest for authentic connection" see chap. 5 in Peters. As Peters suggests, "media—as things that come in between—are liminal objects par excellence, and they deal not only with information but with birth, sex, love and death" (p. 147).

9   *Ikonos* satellite images are owned and sold by the private corporation Space Imaging. Spokespersons for the company appeared on television news programs to present these images to the world. To commemorate the one-year anniversary of 9/11, the company made a digital archive of pre- and postattack images of the World Trade Center and the Pentagon available online free to the public. Some of the *Ikonos* images were used by artist Laura Kurgan in her installation "New York, September 11, 2001 four days later."

10 Wolf Blitzer, of CNN, went so far as to suggest that protests of war after 9/11 were unpatriotic.

11 Foucault, *The Archaeology of Knowledge*, p. 91.

12 Donna Haraway identifies such a position in her essay "Situated Knowledges," in *Simians, Cyborgs, and Women*. For a discussion of modest witnessing see her *Modest_Witness@ Second_Millennium*.

13 Foucault, *The Archaeology of Knowledge*, pp. 83, 89.

14 For further discussion of GPS see my essay "Plotting the Personal." An expanded version of the essay appears in German in the catalog for the Geography and the Politics of Mobility art exhibition, Generali Foundation Gallery, Vienna, Austria, 2003.

15 This is especially the case for scholars in the humanities, whose research on technology often emphasizes local or national sociohistorical contextualization.

16 Morley, *Home Territories*, p. 153.

17 For Kaplan "travel refers not so much to the movements of individuals in the modern era but to the construction of categories in criticism that engender specific ideas and practices" (Kaplan, *Questions of Travel*, p. 2).

18 James Clifford suggests that "there is no politically innocent methodology for intercultural interpretation" (Clifford, "Traveling Cultures," p. 97).

19 Elsaesser, "Digital Cinema," p. 202.

20 Moores, *Satellite Television and Everyday Life*; Mankekar, *Screen Culture, View-*

ing Politics. For a discussion of satellite television audience research see Morley, *Home Territories,* pp. 149–170.

21 Foucault, *The Archaeology of Knowledge,* p. 202.

22 Robbins, *Feeling Global,* pp. 2, 3.

23 Quoted in Williams, "'Perhaps Images at One with the World Are Already Lost Forever,'" p. 381.

24 Virilio, *Strategy of Deception,* chap. 1.

Bibliography

Print Sources

"Aboriginal Map Acquisitions." *Gateways: National Library of Australia*, no. 49 (Feb. 2001).

Adler, Jerry. "Witness at the Creation." *Newsweek*, Nov. 13, 1995, p. 71.

"Alexandria Born Great." *Economist*, June 6, 1998, p. 82.

Allman, William F. "Finding Lost Worlds." *U.S. News and World Report*, Sept. 13, 1993, p. 69.

Anderson, Benedict. *Imagined Communities: Reflections on the Origin and Spread of Nationalism* (London: Verso, 1991).

Anselmo, Joseph C. "Remote Sensing to Alter TV News." *Aviation Week and Space Technology*, Dec. 5, 1994, p.61.

Appadurai, Arjun. "Disjuncture and Difference in the Global Cultural Economy." *Public Culture* 2, no. 2 (1990), pp. 1–24.

———. *Modernity at Large: Cultural Dimensions of Globalization* (Minneapolis: University of Minnesota Press, 1996).

Appleman, Philip. *The Silent Explosion* (Boston: Beacon Press, 1965).

Arden, Harvey. *Dreamkeepers: A Spirit-Journey into Aboriginal Australia* (New York: HarperCollins, 1994).

Aronowitz, Stanley, Barbara Martinsons, and Michael Menser, eds. *Technoscience and Cyberculture: A Cultural Study* (New York: Routledge, 1995).

"Australia's First Aboriginal TV Opens." The Xinhua General Overseas News Service, Jan. 16, 1988.

"Babies Are Focal Point." *Our World* press release, May 1967. NET Collection, National Public Broadcasting Archives, Hornbake Library, University of Maryland, College Park.

Bainbridge, William Sims. "Religions for a Galactic Civilization." In *Science Fiction and Space Futures: Past and Present*, ed. Eugene M. Emme (San Diego, Calif.: American Astronautical Society, 1982), pp. 187–202.

Barbree, Jay. *Destination Mars: In Art, Myth and Science* (New York: Penguin Studio, 1997).

Barthes, Roland. *Mythologies*. Trans. Annette Lavers (New York: Hill and Wang, 1987).

Batchen, Geoffrey. "Da(r)ta." *Afterimage* 24, no. 6 (May/June 1997), pp. 5–6.

Batty, Philip. "Singing the Electric: Aboriginal Television in Australia." In Dowmunt, *Channels of Resistance*, pp. 106–125.

Baudrillard, Jean. "No Pity for Sarajevo." In *This Time We Knew: Western Responses to Genocide in Bosnia*, ed. Thomas Cushman and Stjepan G. Mestrovic (New York: New York University Press, 1996).

———. *Simulations* (New York: Semiotext[e], 1983).

Beatson, Jim. "Dreamtime Beam; Television for Australian Aborigines." *South Magazine*, April 1990, p. 79.

Begley, Sharon. "Out of Egypt, Greece." *Newsweek*, Sept. 23, 1991, p. 49.

———. "Postcards from the Edge." *Newsweek*, Dec. 19, 1994, p. 54.

———."When Galaxies Collide." *Newsweek*, Nov. 3, 1997, pp. 30–37.

Bellamy, Robert V., Jr., and James R. Walker, *Television and Remote Control: Grazing on a Vast Wasteland* (New York: Guilford, 1996).

Benjamin, Walter. *Illuminations*. Trans. Harry Zohn; ed. Hannah Arendt (New York: Harcourt, Brace and World, 1968).

Berko, Lili. "Surveying the Surveilled: Video, Space and Subjectivity." *Quarterly Review of Film and Video* 14, nos. 1/2 (1992), pp. 61–91.

Berland, Jody. "Mapping Space: Imagining Technologies and the Planetary Body." In Aronowitz et al., *Technoscience and Cyberculture*, pp. 123–139.

Berman, Marshall. *All That Is Solid Melts into Air: The Experience of Modernity* (New York: Penguin, 1988).

Bernal, Martin. *Black Athena* (New Brunswick, N.J.: Rutgers University Press, 1987).

Bhabha, Homi K. *The Location of Culture* (London: Routledge, 1994).

"Big Changes Ahead for TV: Here's What You Can Expect." *U.S. News and World Report*, Aug. 6, 1962, p. 80.

"The Bird." *New Yorker*, May 15, 1965, p. 46.

"The Black TV Moguls with Carte Blanche in the Outback." *Australian Financial Review*, Aug. 29, 1986.

Boeke, Kees. *Cosmic View: The Universe in 40 Jumps* (New York: J. Day, 1957).

Bordieu, Pierre. *Distinction: A Social Critique of Judgment and Taste* (Cambridge, Mass.: Harvard University Press, 1987).

"Bosnia-Herzegovina: To Bury My Brothers' Bones." *Amnesty International Country Report*, Amnesty International, EUR 63, no. 15 (1996), pp. 1–22.

"Bosnian Muslim Refugees Visit Homes in Srebrenica Area." *Bosnia Today*, April 28, 2001.

Braidotti, Rosi. *Nomadic Subjects: Embodiment and Sexual Difference in Contemporary Feminist Theory* (New York: Columbia University Press, 1994).

Brodsky, Jack, and Nathan Weiss. *The Cleopatra Papers* (New York: Simon and Schuster, 1963).

Brown, Stuart F. "America's First Eyes in Space." *Popular Science*, Feb. 1996, pp. 42–47.

Browne, Donald R. "Aboriginal Radio in Australia: From Dream Time to Prime Time?" *Journal of Communication* 40, no. 1 (winter 1990), pp. 111–120.

Brunsdon, Charlotte. "Satellite Dishes and the Landscapes of Taste." *New Formations* 15 (winter 1991), pp. 23–42.

Bryld, Mette, and Nina Lykke. *Cosmodolphins: Feminist Cultural Studies of Technology, Animals and the Sacred* (London: Zed Books, 2000).

Buck-Morss, Susan. "Art in the Age of Its Electronic Production." In *Ground Control: Technology and Utopia*, ed. Susan Buck-Morss, Julian Stallabrass, and Leonidas Donskis (London: Black Dog Publishing, 1997).

Buell, Frederick. *National Culture and the New Global System* (Baltimore: Johns Hopkins University Press, 1994).

Burnett, Ron. *Cultures of Vision: Images, Media and the Imaginary* (Bloomington: Indiana University Press, 1995).

Butler, Judith. *Bodies That Matter* (New York: Routledge, 1993).

Butricia, Andrew J., ed. *Beyond the Ionosphere: Fifty Years of Satellite Communication* (Washington: NASA History Office, 1997).

Caldwell, John Thornton. *Televisuality: Style, Crisis, and Authority in American Television* (New Brunswick, N.J.: Rutgers University Press, 1995).

Callaghan, Dympna. "Representing Cleopatra in the Postcolonial Moment." In *Shakespeare's Cleopatra*, ed. Nigel Woods (Buckingham: Open University Press, 1996), pp. 40–65.

Canclini, Nestor Garcia. *Hybrid Cultures: Strategies for Entering and Leaving Modernity* (Minneapolis: University of Minnesota Press, 1995).

Capaccio, Tony, and Mary Greczyn. "Warfare in the Information Age: Military Technology." *Popular Science*, July 1996, p. 52.

Carey, James. *Communication as Culture: Essays on Media and Society* (Boston: Unwin Hyman, 1988).

Chaisson, Eric J. *Hubble Wars: Astrophysics Meets Astropolitics in a Two-Billion-Dollar Struggle over the Hubble Space Telescope* (New York: HarperCollins, 1994).

Chambers, Iain. *Migrancy, Culture, Identity* (London: Routledge, 1994).

Clarke, Arthur C. "Everybody in Instant Touch." *Life*, Sept. 25, 1964, pp. 118–129.

———. "Extra-Terrestrial Relays: Can Rocket Stations Give Worldwide Radio Coverage?" *Wireless World*, Oct. 1945, p. 305.

———. "The World of the Communications Satellite." *UNESCO Courier*, Nov. 1966.

Clifford, James. "Diasporas." *Cultural Anthropology* 9, no. 3 (1994), pp. 302–338.

———. "Traveling Cultures." In *Cultural Studies*, ed. Lawrence Grossberg, Cary Nelson, and Paula A. Treichler (New York: Routledge, 1994), pp. 96–112.

Clifford, James, and George Marcus. *Writing Culture* (Berkeley: University of California Press, 1986).

Cloud, J. G., and K. C. Clarke. "Through a Shutter Darkly: The Tangled Relationships between Civilian, Military, and Intelligence Remote Sensing in the Early U.S. Space Program." In *Secrecy and Knowledge Production*, ed. Judith Reppy (Ithaca, N.Y.: Cornell Peace Studies Program, 1999), pp. 36–56.

Cocca, Aldo Armando. "Consent, Content, Spillover and Participation in Direct Broadcasting from Satellites." *Studies in Broadcasting* 13 (1977), pp. 33–54.

"Communicating by Satellite." *Business Week*, Oct. 27, 1962, p. 98.

Condon, Chris. "Around the World in Two Hours Flat." *National Catholic Reporter*, July 12, 1967.

"Conflict Splits World Telecast." *Broadcasting*, June 26, 1967.

Contractor, Noshir S., A. Singhal, and E. M. Rogers, "Metatheoretical Perspectives on Satellite Television and Development in India." *Journal of Broadcasting and Electronic Media* 32, no. 2 (spring 1988), pp. 129–148.

Corner, John. *Critical Ideas in Television Studies* (Oxford: Oxford University Press, 1999).

Corwin, Phillip. *Dubious Mandate: A Memoir of the UN in Bosnia* (Durham, N.C.: Duke University Press, 1999).

Couldry, Nick, and Anna McCarthy, eds. *Media Space* (London: Routledge, 2003).

Covault, Craig. "Athena Rockets U.S. Back to the Moon." *Aviation Week and Space Technology*, Jan. 12, 1998.

Cowen, Ron. "After Hubble: The Next Generation." *Science News*, April 26, 1997.

———. "Bullies of the Universe: Massive Stars Rob Their Smaller Neighbors." *Science News*, November 30, 1996.

———. "Hubble Views Stellar EGGS." *Science News*, Nov. 4, 1995.

———. "Hubble: A Universe without End." *Science News*, Feb. 1, 1992.

Coyle, Donald. "TV's Global Future Awaits Those Who Take the Opportunity Today." *Variety*, Jan. 4, 1967.

Crary, Jonathan. *Techniques of the Observer: On Vision and Modernity in the Nineteenth Century* (Cambridge, Mass.: MIT Press, 1990).

Crawley, Tony. *The Films of Sophia Loren* (London: LSP Books, 1974).

Curtin, Michael. "Beyond the Vast Wasteland: The Policy Discourse of Global Television and the Politics of American Empire." *Journal of Broadcasting and Electronic Media* 37, no. 2 (spring, 1993), pp. 127–145.

———. *Redeeming the Wasteland: Television Documentary and Cold War Politics* (New Brunswick, N.J.: Rutgers University Press, 1995).

Curtis, Anthony R., ed. *Space Satellite Handbook*, 3rd ed. (Houston: Gulf Publishing, 1994).

"*Dallas* in the Outback." United Press International, Aug. 23, 1986.

Day, Dwayne, John M. Logsdon, and Brian Latell, eds. *Eye in the Sky: The Story of the Corona Spy Satellites* (Washington: Smithsonian Institution Press, 1998).

de Certeau, Michel. *Heterologies: Discourse on the Other*. Trans. Brian Massumi (Minneapolis: University of Minnesota Press, 1986).

———. *The Practice of Everyday Life*. Trans. Steven Rendall (Berkeley: University of California Press, 1988).

———. *The Writing of History* (New York: Columbia University Press, 1988).

Dean, Jodi. *Aliens in America: Conspiracy Cultures from Outerspace to Cyberspace* (Ithaca, N.Y.: Cornell University Press, 1998).

de la Garde, Roger. "Is There a Market for Foreign Cultures?" *Media, Culture and Society* 9 (1987).

Delatiner, Barbara. "Global Show Displays Electronic Magic." *New York Newsday*, June 26, 1967. NET Collection, Series 6, Box 7, folder 9, Wisconsin State Historical Society.

Dickenson, Nicole. "Asia: Supplement-Worldwide Advertising; The Explosion of Asian Satellite TV." *Campaign*, June 7, 1996.

Dirlik, Arlif. "The Postcolonial Aura: Third World Criticism in the Age of Global Capitalism." In *Dangerous Liaisons: Gender, Nation and Postcolonial Perspectives*, ed. Anne McClintock, Aamir Mutti, Ella Shohat, Social Text Collective (Minneapolis: University of Minnesota Press, 1997), pp. 501–528.

Dobbs, Michael, and R. Jeffrey Smith. "New Proof Offered of Serb Atrocities: U.S. Analysts Identify More Mass Grave Sites." *Washington Post*, Oct. 30, 1995.

Dorsey, Gary. *Silicon Sky: How One Small Start Up Went over the Top to Beat the Big Boys into Satellite Heaven* (Reading, Mass.: Perseus Books, 1999).

Douglas, Susan. *Inventing American Broadcasting, 1899–1922* (Baltimore: Johns Hopkins University Press, 1987).

Dowmunt, Tony, ed. *Channels of Resistance: Global Television and Local Empowerment* (London: British Film Institute, 1993).

Durrell, Laurence. *The Alexandria Quartet* (London: R. Maclehose, 1962).

"Early Bird Brightens the Fine Art Scene." *Business Week*, May 29, 1965.

"Early Bird Satellite to Carry Irish Sweeps Derby to ABC-TV Viewers from the Curragh, June 26," ABC Sports press release, June 17, 1965. William Henry Papers, Box 15, "Communications Satellite Corp. 1965" folder, Wisconsin State Historical Society.

"Editorial: The Tokyo Olympics." *New York Times*, Oct. 24, 1964.

Edwards, Paul. *The Closed World: Computers and the Politics of Discourse in Cold War America* (Cambridge, Mass.: MIT Press, 1996).

Ehrlich, Paul R. *The Population Bomb* (New York: Ballantine Books, 1968).

Elsaesser, Thomas. "Digital Cinema: Delivery, Event, Time." In *Cinema Futures: Cain, Abel or Cable? The Screen Arts in the Digital Age*, ed. Thomas Elsaesser and Kay Hoffman (Amsterdam: Amsterdam University Press, 1998), pp. 201–222.

Elson, John. "Attacking Afrocentrism; A Classics Scholar Sharply Challenges the Emerging Theory That Ancient Greece 'Stole' Its Best Ideas from Ancient Egypt." *Time*, Feb. 19, 1996.

Esteinou, Javier. "The Morelos Satellite System and Its Impact on Mexican Society." *Media, Culture and Society* 10 (1988), pp. 419–446.

Estess, Roy S. (Director of NASA Stennis Space Center). Letter to Honorable Trent Lott (U.S. Senate), May 7, 1990. "Archaeology" file, NASA History Office Archive, Washington.

Fair, Jo Ellen. "Francophonie and the National Airwaves: A History of Television in Senegal." In Parks and Kumar, *Planet TV*, pp. 189–210.

Fair, Jo Ellen, and Lisa Parks. "Africa on Camera: Televised Video Footage and Aerial Imaging of the Rwandan Refugee Crisis." *Africa Today* 48 (2001), pp. 35–58.

Featherstone, Mike, ed. *Global Culture: Nationalism, Globalization and Modernity* (London: Sage, 1990).

Feenberg, Andrew. *Questioning Technology* (London: Routledge, 1999).

Ferguson, Douglas A. "Channel Repertoire in the Presence of Remote Control Devices, VCRs, and Cable Television." *Journal of Broadcasting and Electronic Media* (winter 1992).

Feuer, Jane. "The Concept of Live Television: Ontology as Ideology." In *Regarding*

Television: Critical Approaches—An Anthology, ed. E. Ann Kaplan (Frederick, Md.: University Publications of America, 1983), pp. 12–22.

Fiske, John. *Media Matters: Everyday Culture and Political Change* (Minneapolis: University of Minnesota Press, 1994).

———. *Power Plays, Power Works* (London: Verso, 1993).

Foucault, Michel. *The Archaeology of Knowledge.* Trans. A. M. Sheridan Smith (New York: Harper and Row, 1976).

———. *Discipline and Punish: The Birth of the Prison* (New York: Vintage, 1977).

———. "Nietzsche, Genealogy, History." In *The Foucault Reader*, ed. Paul Rabinow (London: Penguin, 1984), pp. 76–100.

———. *Power/Knowledge: Selected Interviews and Other Writings 1972–1977.* Ed. Colin Gordon (New York: Pantheon Books, 1980).

Fowler, Gene, and Bill Crawford. *Border Blasters* (Austin: Texas Monthly Press, 1987).

Fraser, Nancy. "Rethinking the Public Sphere: A Contribution to the Critique of Actually Existing Democracy." In *Social-Text* 25–26 (1990), pp. 56–80.

Friedberg, Anne. *Window Shopping: Cinema and the Postmodern* (Berkeley: University of California Press, 1994).

Gandy, Oscar. *The Panoptic Sort: A Political Economy of Personal Information* (Boulder, Colo.: Westview Press, 1993).

Gibbons, Ann. "A 'New Look' for Archaeology: New High-Tech Methods for Finding Fossils from the Air Are Becoming a Standard Part of the Archaeologist's Took Kit." *American Association for the Advancement of Science*, May 17, 1991, p. 918.

Giddens, Anthony. *Beyond Left and Right: The Future of Radical Politics* (Palo Alto: Stanford University Press, 1995).

———. *The Nation-State and Violence* (Cambridge, U.K.: Polity, 1987).

Ginsberg, Faye. "Embedded Aesthetics: Creating a Discursive Space for Indigenous Media." *Cultural Anthropology* 9, no. 3 (1994), pp. 365–382.

Glynn, Kevin. *Tabloid Culture: Trash Taste, Popular Power, and the Transformation of American Television* (Durham, N.C.: Duke University Press, 2000).

Goldberg, Ken, ed. *The Robot in the Garden: Telerobotics and Telepistemology in the Age of the Internet* (Cambridge, Mass.: MIT Press, 2000).

Gomez-Peña, Guillermo. *New World Border: Prophecies, Poems and Loqueras for the End of the Century* (San Francisco: City Lights Books, 1996).

Gonzalez, Jennifer. "Autotopographies." In *Prosthetic Territories: Politics and Hypertechnologies*, ed. Gabriel Brahm Jr. and Mark Driscoll (Boulder, Colo.: Westview Press, 1995) pp. 133–150.

Gow, James, Richard Paterson, and Alison Preston, eds. *Bosnia by Television* (London: British Film Institute, 1997).

Graff, James. "Commanders in Court: The Trial Begins of the Man Accused of Overseeing the Worst Crime in Europe since World War II." *Time Europe*, March 27, 2000.

Grewal, Inderpal. *Home and Harem: Nation, Gender, Empire, and the Cultures of Travel* (Durham, N.C.: Duke University Press, 1996).

Griffiths, Alison. " 'To the World the World We Show': Early Travelogues as Filmed Ethnography." *Film History* 11 (1999), pp. 282–307.

Gundle, Stephen. "Sophia Loren, Italian Icon." *Historical Journal of Film, Radio and Television* 15, no. 3 (1995).

Hall, Alan. "Slices of the Past." *Scientific American*, June 22, 1998.

Hamer, Mary. *Signs of Cleopatra: History, Politics, Representation* (London: Routledge, 1993).

Haraway, Donna. *Modest_Witness@ Second_Millennium.FemaleMan©_Meets_ OncoMouse: Feminism and Technoscience™* (New York: Routledge, 1997).

———. *Simians, Cyborgs, and Women: The Reinvention of Nature* (New York: Routledge, 1991).

Harding, Sandra. *Whose Science? Whose Knowledge? Thinking from Women's Lives* (Ithaca, N.Y.: Cornell University Press, 1991).

Harrison, David. *The Sociology of Modernization and Development* (London: Unwin Hyman, 1988).

Hartley, John. *The Uses of Television* (London: Routledge, 2000).

Hartouni, Valerie. *Cultural Conceptions: On Reproductive Technologies and the Remaking of Life* (Minneapolis: University of Minnesota Press, 1997).

Hauser, Philip. *The Population Dilemma* (Englewood Cliffs, N.J.: Prentice-Hall, 1963).

Hawking, Stephen W. *A Brief History of Time: From the Big Bang to Black Holes* (New York: Bantam, 1998).

Hayles, N. Katherine. "Narratives of Artificial Life." In *Future Natural*, ed. George Robertson, Melinda Mash, Lisa Tickner, Jon Bird, Barry Curtis, and Tim Putnam (New York: Routledge, 1996), pp. 146–164.

Hayward, Philip, and Tana Wollen, eds. *Future Vision: New Technologies of the Screen* (London: British Film Institute, 1993).

Hebidge, Dick. *Hiding in the Light* (London and New York: Routledge, 1988).

Higashi, Sumiko. "Antimodernism as Historical Representation in a Consumer Culture: Cecile B. DeMille's *The Ten Commandments*, 1923, 1956, 1993." In Sobchack, *The Persistence of History*, pp. 91–112.

"High Tech for the Bush Aborigines." *Australian Financial Review*, June 19 1989.

Higham, Charles. *Cecil B. DeMille* (New York: Charles Scribner's Sons, 1973).

Hodge, Robert. "Aboriginal Truth and White Media: Eric Michaels Meets the Spirit of Aboriginalism." *Continuum: The Australian Journal of Media and Culture* 3, no. 2 (1990).

Holland, Barbara. "Cleopatra: What Kind of Woman Was She Anyway?" *Smithsonian*, Feb. 1997, p. 56.

Honig, Jan Willem, and Norbert Both. *Srebrenica: Record of a War Crime* (New York: Penguin, 1996).

Hotz, Robert Lee. "Comet Fragments Begin Crashing to Jupiter." *Los Angeles Times*, July 17, 1994, p. AI.

———. "The Impact of Cosmic Collisions." *Los Angeles Times*, July 11, 1994, p. AI.

Hough, Harold. *Satellite Surveillance* (Port Townsend, Wash.: Loompanics Unlimited, 1991).

"How People in Europe Reacted to Telstar." *U.S. News and World Report,* July 23, 1962, p. 38.

Hozic, Aida A. "Making of the Unwanted Colonies: (Un)imagining Desire." In *Cultural Studies and Political Theory,* ed. Jodi Dean (Ithaca, N.Y.: Cornell University Press, 2000), pp. 228–243.

Hughes-Hallet, Lucy. *Cleopatra: Histories, Dreams and Distortions* (New York: Harper and Row, 1990).

Huntington, Tom. "The Whole World's Watching." *Air and Space,* May 1996, pp. 55–60.

Huseonica, Susan. "Archaeology from Above." NASA *Magazine,* summer 1993, pp. 17–20.

Ignatieff, Michael. *Virtual War: Kosovo and Beyond* (New York: Picador USA, 2000).

"Imparja: Keeping Strong" press kit (Alice Springs: Imparja Television, 1998).

"Inaugural Ceremonies to Mark Commercial Debut of Early Bird Monday," Communications Satellite Corporation press release, June 25, 1965. William Henry Papers, Box 15, "Communications Satellite Corp. 1965" folder, Wisconsin State Historical Society.

"Interstellar Migration and the Population Problem." *Heredity* 50, nos. 68–70 (1959).

Jameson, Fredric, and Masao Miyoshi, eds. *The Cultures of Globalization* (Durham, N.C.: Duke University Press, 1998).

Jennings, Karen. *Sites of Difference: Cinematic Representations of Aboriginality and Gender* (South Melbourne: Australian Film Institute, 1993).

"Journey to the Beginning of Time." *U.S. News and World Report,* March 25, 1990, p. 52.

Joyce, Christopher. "Archaeology Takes to the Skies." *New Scientist,* Jan. 25, 1992, pp. 42–46.

Kaplan, Caren. *Questions of Travel: Postmodern Discourses of Displacement* (Durham, N.C.: Duke University Press, 1996).

Keizer, Gregg. "Stones on the Screen: Use of Computer-Aided Design by Archaeologists." *Omni,* Nov. 1992, p. 8.

Keller, Evelyn Fox. *A Feeling for the Organism* (New York: Freeman, 1985).

Keller, Evelyn Fox, and Carol Grontowsky. "The Mind's Eye." In *Discovering Reality,* ed. Sandra Harding and Mary Hinitikka (Reidel: Dordrecht, 1983).

Kellner, Douglas. *The Persian Gulf TV War* (Boulder, Colo.: Westview Press, 1992).

Kember, Sarah. "Feminist Figuration and the Question of Origin." In *Future Natural: Nature, Science, Culture,* ed. George Robertson, Melinda Mash, Lisa Tickner, Jon Bird, Barry Curtis, and Tim Putnam (London: Routledge, 1996), pp. 265–269.

Kinsley, Michael E. *Outer Space and Inner Sanctums: Government, Business and Satellite Communication* (New York: Wiley, 1976).

Koch, Howard. *The Panic Broadcast: Portrait of an Event* (Boston: Little, Brown, 1970).

Landay, Jonathan S. "US Troops in Bosnia Depend on Array of High-Tech Eyes, Ears." *Christian Science Monitor,* Dec. 26, 1996, p. 1.

Langton, Marcia. *"Well, I Heard It on the Radio and I Saw It on the Television": An Essay for the Australian Film Commission on the Politics and Aesthetics of Film-making by and about Aboriginal People and Things* (Sydney: Australian Film Commission, 1993).

Lavin, Chris. "An Educated Guess." *St. Petersburg Times*, July 12, 1994, p. 3D.

Lee, Mary Ann. " 'Our World' Sees Advent of New Eras in Television." *Memphis Press Scimitar*, June 26, 1967. NET Collection, Series 6, Box 7, folder 9, Wisconsin State Historical Society.

Lefkowitz, Mary. *Not Out Of Africa: How Afrocentrism Became an Excuse to Teach Myth as History* (New York: Basic Books, 1996).

———. "Not Out of Africa: The Origins of Greece and the Illusions of Afrocentrists." *New Republic*, Feb. 10, 1992, p. 29.

Leigh, Michael. "Curiouser and Curiouser." In *Back of Beyond: Discovering Australian Film and Television*, ed. Scott Murray (Sydney: Australian Film Commission, 1988).

Lemonick, Michael. "Cosmic Close-Ups." *Time*, Nov. 25, 1995, p. 95.

Lenica, Jan, and Alfred Sauvy. *The Population Explosion* (New York: Dell, 1962).

Levy, David. *Comets: Creators and Destroyers* (New York: Touchstone, 1998).

Levy, Stan. "Memo to All Stations." June 21, 1967. NET Files, Series 6, Box 7, folder 9, NET Collection, Wisconsin State Historical Society.

Liebes, Tamar, and Elihu Katz. "On the Critical Abilities of Television." In *Remote Control: Television, Audiences, and Cultural Power*, ed. Ellen Seiter, Hans Borckers, Gabriele Kreutzner, and Eva Marie Warth (London, New York: Routledge, 1989), pp. 204–222.

Lightfoot, Victoria, and Dale Lightfoot. "Revealing the Ancient World through High Technology: Remote Sensing." *MIT Technology Review* 92, no. 4 (May 1989).

Longinovic, Toma. "Music Wars: Blood and Song at the End of Yugoslavia." In *Music and the Racial Imagination*, ed. Ronald Radano and Philip Bohlman (Chicago: University of Chicago Press, 2000), pp. 622–643.

———. "Vampires like Us: Gothic Imaginary and 'the Serbs.' " In *Balkan as Metaphor: Between Globalization and Fragmentation*, ed. Dusan I. Bjelic and Obrad Savic (Cambridge, Mass.: MIT Press, 2002), pp. 39–59.

Lowe, Lisa. *Immigrant Acts: On Asian American Cultural Politics* (Durham, N.C.: Duke University Press, 1996).

Lunenfeld, Peter, ed. *The Digital Dialectic* (Cambridge, Mass: MIT Press, 1999).

Luther, Sara Fletcher. *The United States and the Direct Broadcast Satellite: The Politics of International Broadcasting in Space* (New York: Oxford University Press, 1988).

Lutz, Catherine, and Jane Collins. *Reading "National Geographic"* (Chicago: University of Chicago Press, 1993).

Lyon, David. *The Electronic Eye: The Rise of Surveillance Society* (Minneapolis: University of Minnesota Press, 1994).

Mack, Pamela. *Viewing the Earth: The Social Construction of the Landsat Satellite System* (Cambridge, Mass.: MIT Press, 1990).

MacKenzie, Donald, and Judy Wajcman, eds. *The Social Shaping of Technology* (Milton Keynes: Open University Press, 1985).

Malone, Peter. *In Black and White and Colour: A Survey of Aborigines in Australian Feature Films* (Victoria: Spectrum Publications, 1987).

Mankekar, Purnima. *Screen Culture, Viewing Politics: An Ethnography of Television, Womanhood and Nation in Postcolonial India* (Durham, N.C.: Duke University Press, 1999).

Manovich, Lev. *The Language of New Media* (Cambridge, Mass.: MIT Press, 2001).

Marks, Laura. *Touch: Sensuous Theory and Multisensory Media* (Minneapolis: University of Minnesota Press, 2002).

Marvin, Carolyn. *When Old Technologies Were New: Thinking about Electric Communication in the Late Nineteenth Century* (New York: Oxford University Press, 1988).

Mattelart, Armand. *Mapping World Communication: War, Progress, Culture* (Minneapolis: University of Minnesota Press, 1994).

May, Stephen. "The Object at Hand: The Circuitous Route of Edmonia Lewis' Masterwork." *Smithsonian*, Sept. 1996.

McAleer, Neil. "Tom Sever: Archaeologist and Remote Sensing Scientist." *Omni*, Feb. 1994.

McAnany, Emile and Joao Batista Oliverira, *The SACI/EXERN Project in Brazil: An Analytical Case Study* (New York: Unesco, 1980).

McCabe, Colin. "Realism and the Cinema: Notes on Some Brechtian Theses." *Screen* 15, no. 2 (1974), pp. 21–27.

McCarthy, Anna. *Ambient Television: Visual Culture and Public Space* (Durham, N.C.: Duke University Press, 2001).

McCarthy, Patrick. *After the Fall: Srebrenica Survivors in St. Louis* (St. Louis: Missouri Historical Society, 2000).

McClintock, Anne. *Imperial Leather: Race, Gender and Sexuality in the Colonial Conquest* (New York: Routledge, 1995).

McElroy, John H. "Observing the Earth from Space." *Futurist*, Jan.–Feb. 1986.

Mclear, Kyo. *Beclouded Visions: Hiroshima-Nagasaki and the Art of Witness* (New York: SUNY Press, 1998).

McLuhan, Marshall. *Understanding Media* (New York: McGraw Hill, 1964).

McLuhan, Marshall, and Quentin Fiore. *The Medium Is the Massage* (New York: Bantam, 1967).

Mellencamp, Patricia. *High Anxiety: Catastrophe, Scandal, Age and Comedy* (Bloomington: Indiana University Press, 1992).

Menser, Michael, and Stanley Aronowitz. "On Cultural Studies, Science, and Technology." In Aronowitz et al., *Technoscience and Cyberculture.*

Michaels, Eric. *The Aboriginal Invention of Television, 1982–1986* (Canberra: Australian Institute of Aboriginal Studies, 1986).

———. *Bad Aboriginal Art: Tradition, Media, and Technological Horizons* (Minneapolis: University of Minnesota Press, 1994).

Mills, Mike. "Satellite Glitch Cuts Off Data Flow." *Washington Post*, May 21, 1998.

Mirzoeff, Nicholas. *An Introduction to Visual Culture* (New York: Routledge, 1999).

———, ed. *The Visual Culture Reader*, second edition (London: Routledge, 2002).

Mitchell, Tony. "Treaty Now! Indigenous Music and Music Television in Australia." *Media, Culture and Society* 15 (1993), pp. 299–308.

Mitchell, William J. *The Reconfigured Eye: Visual Truth in the Post-Photographic Era* (Cambridge, Mass: MIT Press, 1992).

Moores, Shaun. *Satellite Television and Everyday Life* (Luton: University of Luton Press, 1996).

———. "Satellite TV as Cultural Sign: Consumption, Embedding and Articulation." *Media, Culture and Society* 15, no. 4 (Oct. 1993), pp. 621–639.

Morley, David. *Home Territories: Media, Mobility and Identity* (London: Routledge, 2000).

Morley, David, and Kevin Robins. *Spaces of Identity: Global Media, Electronic Landscapes and Cultural Boundaries* (London: Routledge, 1995).

Morse, Margaret. *Virtualities: Television, Media Art and Cyberculture* (Bloomington: Indiana University Press, 1998).

Morsy, Soheir. "Biotechnology and the Taming of Women's Bodies." In *Processed Lives: Gender and Technology in Everyday Life*. ed. Jennifer Terry and Melodie Calvert (London: Routledge, 1997), pp. 165–173.

Mulhern, Francis. "*Towards 2000*, or News from You-Know-Where." In *Raymond Williams: Critical Perspectives*, ed. Terry Eagleton (Boston: Northeastern University Press, 1989), pp. 5–30.

Mulvey, Laura. "Visual Pleasure and Narrative Cinema." In *Issues in Feminist Film Criticism*, ed. Patricia Erens (Bloomington: Indiana University Press, 1990), pp. 28–40.

Naeye, Robert. *Through the Eyes of Hubble: The Birth, Life, and Violent Death of Stars* (Waukesha, Wis.: Kalmback Publishing, 1998).

Naficy, Hamid, ed. *Home Exile, Homeland: Film, Media and the Politics of Place* (London: Routledge, 1999).

Nakamura, Lisa. " 'Where Do You Want to Go Today?' Cybernetic Tourism, the Internet and Transnationality." In *Race in Cyberspace*, ed. Beth E. Kolko, Lisa Nakamura, and Gilbert B. Rodman (New York: Routledge, 2000), pp. 15–26.

"NBC's Olympic Hurdle." *Business Week*, Oct. 10, 1964, p. 34.

Negrine, Ralph, ed. *Satellite Broadcasting: The Politics and Implications of New Media* (London: Routledge, 1988).

Newcomb, Horace, and Paul M. Hirsch. "Television as a Cultural Forum." in *Television: The Critical View*, ed. Horace Newcomb, 5th ed. (Oxford: Oxford University Press, 1994), pp. 503–515.

———. "Toward a Television Aesthetic." In *Television: The Critical View*, ed. Horace Newcomb, 4th ed. (Oxford: Oxford University Press, 1987), pp. 611–627.

Newcott, William R. "Time Exposures." *National Geographic*, April 1997, p. 3.

Notestein, Frank W., Dudley Kirk, and Sheldon Segal. "The Problem of Population Control." In *The Population Dilemma*, ed. Philip Hauser, 2nd ed. (Englewood Cliffs, N.J.: Prentice-Hall, 1969), pp. 139–167.

O'Connell, Sanjida. "Unearthing Bodies of Evidence: Forensic Software Is Helping to Identify Bosnia's Victims of Slaughter." *The Guardian*, July 25, 1996, p. 8.

O'Regan, Tom. *Australian Television Culture* (St. Leonards: Allen and Unwin, 1993).

————. "TV as Cultural Technology: The Work of Eric Michaels." *Continuum: The Australian Journal of Media and Culture* 3, no. 2 (1990).

O'Toole, Fintan. "The Dredd of 2000 AD." *The Guardian*, Jan. 7, 1995, p. 29.

Our World press release. June 25, 1967. NET Files, Series 6, Box 7, folder 9, NET Collection, Wisconsin State Historical Society.

"'Our World'—Round the World Project Fact Book." May 1, 1967. NET Collection, National Public Broadcasting Archives, Hornbake Library, University of Maryland, College Park.

Our World script. June 1967. NET Collection, National Public Broadcasting Archives, Hornbake Library, University of Maryland, College Park.

Panek, Richard. *Seeing and Believing: How the Telescope Opened Our Eyes and Minds to the Heavens* (New York: Penguin, 1998).

Parapak, Jonathan L. "The Role of Satellite Communications in National Development." *Media Asia* 20, no. 1 (1993), pp. 24–32.

Parenti, Christian. "Crossing Borders: The Police, Military and Border Patrol Have Joined Forces to Crack Down on Drugs and Illegal Immigrants." *In These Times*, March 22, 1998, p. 15.

Parenti, Michael. *To Kill a Nation: The Attack on Yugoslavia* (London: Verso, 2001).

Parks, Lisa. "As the Earth Spins: NBC's *Wide Wide World* and Early Live Global Television." *Screen* 42, no. 4 (winter 2001), pp. 332–349.

————. "Plotting the Personal: Global Positioning Satellites and Interactive Media." *Ecumene: A Journal of Cultural Geographies* 8, no. 2 (2001), pp. 209–222.

————. "Technology in the Twilight: A Cultural History of the First Earth Satellite." *Humanities and Technology Review* 16 (fall 1997), pp. 3–20.

Parks, Lisa, and Shanti Kumar, eds. *Planet TV: A Global Television Reader* (New York: New York University Press, 2002).

Parry, Benita. "Problems in Current Theories of Colonial Discourse." in *The Post-Colonial Studies Reader*, ed. Bill Ashcroft, Gareth Griffiths, and Helen Tiffin (London: Routledge, 1995), pp. 36–44.

"Pathfinding in a New Era." *Newsday*, May 20, 1997, p. B29.

Peebles, Curtis L. *The Corona Project: America's First Spy Satellites* (Annapolis: U.S. Naval Institute, 1997).

Pelton, Joseph N. "Project SHARE and the Development of Global Communications." In *Beyond the Ionosphere: Fifty Years of Satellite Communication*, ed. Andrew Butricia (Washington: NASA History Series, 1997), pp. 257–264.

Penley, Constance. *NASA/TREK: Popular Science and Sex in America* (London: Verso, 1997).

————. "Time Travel, Primal Scene and the Critical Dystopia." In *Alien Zone: Cultural Theory and Contemporary Science Fiction Cinema*, ed. Annette Kuhn (London: Verso, 1990), pp. 116–127.

Peters, John Durham. *Speaking into the Air: A History of the Idea of Communication* (Chicago: University of Chicago Press, 1999).

"Photo Cutlines." Our World press packet, May 1967. NET Collection, National Public Broadcasting Archives, Hornbake Library, University of Maryland, College Park.

Poggioli, Sylvia. "Scouts without Compasses." *Nieman Reports*, originally published fall 1993. Reprinted in special issue vol. 53–54, 2000.

Porterfield, Tod. "Egyptomania! Western Art That Is Inspired by Egyptian Art." *Art in America*, Nov. 1994, pp. 84–90.

Potter, Lou. "Memo to All Stations." June 23, 1967. NET Files, Series 6, Box 7, folder 9, NET Collection, Wisconsin State Historical Society.

Pratt, Mary Louise. *Imperial Eyes: Travel Writing and Transculturation* (London: Routledge, 1992).

"Production Philosophy." Undated. NET Collection, National Public Broadcasting Archives, Hornbake Library, University of Maryland, College Park.

"Public Eye: Spy Satellite Technology May Assist Government Watchdogs." *Scientific American*, Aug. 1996.

"Putting Space to Work to Educate the World." *Business Week*, Dec. 25, 1965, p. 17.

Rapaport, Richard. "World War 3.1: The Shape of Things to Come?" *Forbes*, Oct. 7, 1996, p. 125–127, 129.

Redfield, Peter. *Space in the Tropics: From Convicts to Rockets in French Guiana* (Berkeley: University of California Press, 2000).

Reid, Roddey, and Sharon Traweek, eds. *Doing Science + Culture* (New York: Routledge, 2000).

Reston, James, Jr. "Orion: Where Stars Are Born." *National Geographic*, Dec. 1995, pp. 91–92.

Review of *Cleopatra*. *Variety*, Aug. 21, 1934.

Richelson, Jeffrey T., and Danielle Gordon. "High Flyin' Spies." *Bulletin of Atomic Scientists* 52, no. 5 (1996), p. 48.

Robbins, Bruce. *Feeling Global: Internationalism in Distress* (New York: New York University Press, 2000).

Robertson, Roland. *Globalization: Social Theory and Global Culture* (London: Sage, 1992).

Robins, Kevin. *Into the Image: Culture and Politics in the Field of Vision* (London: Routledge, 1996).

Rogoff, Irit. *Terra Infirma: Geography's Visual Culture* (London: Routledge, 2000).

Rohde, David. *Endgame: The Betrayal and Fall of Srebrenica, Europe's Worst Massacre Since World War II* (Boulder: Westview Press, 1997).

"The Room Sized World." *Time*, May 14, 1965, pp. 84–85, 88.

Ross, Andrew. *Strange Weather: Culture, Science and Technology in the Age of Limits* (London: Verso, 1991).

"Round the World Project." Report of Meeting Held in Geneva at the Hotel de la Paix, Feb. 13, 1967. NET Collection, National Public Broadcasting Archives, Hornbake Library, University of Maryland, College Park.

Ryan, Linda. "What's in a Mass Grave?" *Living Marxism*, vol. 88, March 1996.

Sagan, Carl. *Contact: A Novel* (New York: Simon and Schuster, 1985).

Sandoval, Chela. *The Methodology of the Oppressed* (Minneapolis: University of Minnesota Press, 2000).

Sassen, Saskia. *The Global City: New York, London, Tokyo.* 2nd ed. (Princeton, N.J.: Princeton University Press, 2001).

"Satellite Takes Photo of Satan's Face—On Planet Earth!" *Weekly World News,* June 25, 1996, p. 1.

"Satellite to Put Europe on TV." *Popular Science,* June, 1962, p. 73.

Schiller, Dan. *Digital Capitalism: Networking the Global Market System* (Cambridge, Mass.: MIT Press, 1999).

Schiller, Herbert. *Mass Communications and American Empire* (Boulder, Colo.: Westview Press, 1992).

Schneider, Cynthia, and Brian Wallis, eds. *Global Television* (New York: Wedge Press, 1988).

Sconce, Jeffrey. *Haunted Media: Electronic Presence from Telegraphy to Television* (Durham, N.C.: Duke University Press, 2001).

Scott, David Clark. "Joining Together to Build Self Image." *Christian Science Monitor,* Feb. 21, 1989, p. 6.

Seiter, Ellen, Hans Borchers, Gabriele Kreutzner, and Eva Marie Warth, eds. *Remote Control: Television, Audiences and Cultural Power* (London: Routledge, 1989).

"Serbian President, NATO Support Mass Grave Investigators." Associated Press, Jan. 22, 1996.

Shapiro, Michael J., and Hayward R. Alker, eds. *Challenging Boundaries: Global Flows, Territorial Identities* (Minneapolis: University of Minnesota Press, 1996).

Shine, Eric. "Biz Blasts Off." *Business Week,* Jan. 27, 1997, p. 63.

Shoesmith, Brian. "Technology Transfer or Technology Dialogue: Rethinking Western Communication Values." *Media Asia* 20, no. 3 (1993), pp. 152–156.

Shohat, Ella, and Robert Stam. *Unthinking Eurocentrism: Multiculturalism and the Media* (London: Routledge, 1994).

Silverman, Kaja. *The Threshold of the Visible World* (New York: Routledge, 1996).

Smyth, Roseleen. " 'White Australia Has a Black Past': Promoting Aboriginal and Torres Strait Islander Land Rights on Television and Video." *Historical Journal of Film, Radio and Television* 15, no. 1 (1995), pp. 105–123.

Sobchack, Vivian, ed. *The Persistence of History: Cinema, Television, and the Modern Event* (New York: Routledge, 1996).

———. *Screening Space: The American Science Fiction Film* (New York: Ungar, 1987).

Sobel, Dava. "Among Planets." *New Yorker,* Dec. 9, 1996, p. 90.

Spigel, Lynn. "From Domestic Space to Outer Space: The 1960s Fantastic Family Sitcom." In *Close Encounters,* ed. Constance Penley, Elisabeth Lyon, Lynn Spigel, and Janet Bergstrom (Minneapolis: University of Minnesota Press, 1991), pp. 205–236.

———. *Make Room For TV* (Chicago: University of Chicago Press, 1992).

———. *Welcome to the Dreamhouse* (Durham, N.C.: Duke University Press, 2001).

"The Sputnik." *Time,* Oct. 14, 1957, p. 46.

Stabile, Carol. "Shooting the Mother." In Treichler et al., *The Visible Woman,* pp. 171–197.

"Star Searchers." *Washington Post,* April 14, 1996, p. W10.

Steichen, Edward. Introduction to *The Family of Man* (New York: Simon and
 Schuster, 1993).

Stover, Eric, and Gilles Peress. *The Graves: Srebrenica and Vukovar* (Zurich: Scalo
 Publishers, 1998).

Stucker, Hal. "Junkosphere." *Wired*, Feb. 1998, p. 40.

Sudetic, Chuck. *Blood and Vengeance: One Family's Story of the War in Bosnia* (New
 York: Penguin Books, 1999).

"Sunken Treasures: Divers Discover Alexandria's Secret History." *Time For Kids*,
 Nov. 22, 1996, p. 6.

"Suspect Tells of Srebrenica War Crimes." Reuters World Service, Mar. 8, 1996.

Taylor, Penny, ed., *After 200 Years: Photographic Essays of Aboriginal and Islander
 Australia Today* (Cambridge, U.K.: Cambridge University Press, 1988).

"Television Goes Global: Footprints in the Sand." *Economist*, June 18, 1988.

"Telstar's TV Future Is Foggy." *Business Week*, July 7, 1962.

Tirenzi, Fiorella. *Heavenly Knowledge: An Astrophysicist Seeks Wisdom in the Stars*
 (New York: Avon Books, 1998).

Treichler, Paula, Lisa Cartwright, and Constance Penley, eds. *The Visible Woman*
 (New York: New York University Press, 1998).

Trumbull, Robert. "City Gay, Excited with Week to Go." *New York Times*, Oct. 4,
 1964, v:1.

———. "Crowd of 72,000 Hails Big Parade." *New York Times*, Oct. 10, 1964, p. 3.

Tuathail, Gearóid Ó. "Borderless Worlds? Problematizing Discourses of Deterri-
 torialization." In GEOPOLITICS 4, no. 2 (2000).

"TV Gets a Boost through Space." *Business Week*, July 14, 1962, pp. 32–33.

"TV via Satellite: Still Long Way to Go." *U.S. News and World Report*, Dec. 24, 1962.

UNESCO, *Communication in the Space Age: The Use of Satellites by the Mass Media* (The
 Hague: UNESCO, 1968).

"A U.S. First: Transatlantic TV via Space." *U.S. News and World Report*, July 23, 1962.

Valaskakis, Gail. "Communication, Culture and Technology: Satellites and Northern Native Broadcasting in Canada." In *Ethnic Minority Media: An International
 Perspective*, ed. Stephen Harold Riggens (Newbury Park, Calif.: Sage, 1992),
 pp. 63–81.

Van Der Werf, Martin. "Astronomer Finally Going to the Moon—Posthumously."
 Arizona Republic, Jan. 7, 1998, p. A1.

Vesilind, Priit J. "In Focus: Bosnia." *National Geographic*, June 1996, p. 50.

Virilio, Paul. *Lost Dimension* (Cambridge, Mass.: MIT Press, 1991).

———. *Open Sky*. Trans. Julie Rose (London: Verso, 1997).

———. *Strategy of Deception*. Trans. Chris Turner (London: Verso, 2001).

———. *The Vision Machine* (London: British Film Institute, 1994).

———. *War and Cinema: The Logistics of Perception* (London: Verso, 1989).

Vismann, Cornelia. "Starting from Scratch: Concepts of Order in No Man's Land."
 In *War, Violence and the Modern Condition*, ed. Bernd Huppauf (Berlin: Walter de
 Gruyter, 1997), pp. 46–64.

Wajcman, Judy. *Feminism Confronts Technology* (University Park: Pennsylvania State
 University Press, 1991).

Walker, James R., and Robert V. Bellamy Jr., eds. *The Remote Control in the New Age of Television* (Westport, Conn.: Praeger, 1993).

"War Crimes Prosecutor Wants Team to Hunt Suspects." *Bosnia Today*, May 2, 2001.

Wark, McKenzie. *Virtual Geography: Living with Global Media Events* (Bloomington: Indiana University Press, 1994).

Watson, Mary Ann. *The Expanding Vista: American Television in the Kennedy Years* (New York: Oxford University Press, 1990).

Weber, Samuel. *Mass Mediauras: Form, Technics, Media* (Stanford, Calif.: Stanford University Press, 1996).

Weiner, Tim. "U.S. Says Serbs May Have Tried to Destroy Massacre Evidence." *New York Times*, Oct. 30, 1995, p. A8.

Wellborn, Stanley N. "History's Secrets Yielding to High-Tech Devices." *U.S. News and World Report*, Feb. 18, 1985, p. 68.

White, Hayden. "The Modernist Event." In Sobchack, *The Persistence of History*, pp. 17–38.

White, Mimi. "Flow and Other Close Encounters with Television." In Parks and Kumar, *Planet TV*, pp. 94–110.

"Who Art in Heaven?" *Harper's Magazine*, April 1996, p. 23.

Wilford, John Noble. "Scientists Step Up Search for Extrasolar Planets." *New York Times*, Feb. 9, 1997.

Williams, Raymond. *Television: Technology and Cultural Form* (New York: Schocken Books, 1974).

———. *Towards 2000* (London: Chatto and Windus, 1983).

Williams, Sarah. "'Perhaps Images at One with the World Are Already Lost Forever': Visions of Anthropology in Post-Cultural Worlds." In *The Cyborg Handbook*, ed. Chris Hables Gray (New York: Routledge, 1995), pp. 379–392.

Wilson, Jack. Press release regarding "Our World," May 18, 1967. NET Files, Series 6, Box 7, folder 9, NET Collection, Wisconsin State Historical Society.

Wilson, Rob, and Wimal Dissanayake. *Global/Local: Cultural Production and the Transnational Imaginary* (Durham, N.C.: Duke University Press, 1996).

Winston, Brian. *Technologies of Seeing: Photography, Cinematography and Television* (London: British Film Institute, 1996).

Woodward, Susan L. "War: Building States from Nations." In *Masters of the Universe? NATO's Balkan Crusade*, ed. Tariq Ali (London: Verso, 2000), pp. 203–270.

Zumach, Andreas. "US Intelligence Knew Serbs Were Planning an Assault on Srebrenica." *BASIC Reports* 47, no. 16 (1995).

Online Sources

Antic, Miroslav. "Srebrenica Massacre Denial—Defence Witnesses Refuse to Accept Massacres Took Place." Anti-Imperialist League, Nov. 17, 2000. http://www.lai-aib.org/lai/article_lai.phtml?section=A3ABBBAA&object_id=2191 (accessed June 4, 2001; printouts of page in author's possession).

Barela, Timothy P. "Given' em Their Space: Peacekeeping Forces in Bosnia

Rely on 'High-Ground' Support." *Airman Magazine,* June 1996. Available at
http://www.af.mil/news/airman/0696/givespac.htm, accessed Aug. 20, 2004.

"The Bosnia Massacre." Public Eye Project, Federation of American Scientists.
http://www.fas.org/eye/bosnia.htm (accessed Aug. 7, 1998; printouts of page in
author's possession).

Buchan, Glenn. "Information War and the Air Force: Wave of the Future? Current
Fad?" Issue Paper, Project Air Force, RAND, March 1996. http://www.rand.org./
publications/IP/IP149 (accessed Aug. 1, 1998; printouts of page in author's
possession).

Buchtmann, Lydia. "Digital Songlines: The Use of Modern Communication Technology by an Aboriginal Community in Remote Australia." Report for Department of Communications, Information Technology and the Arts, Australia,
1999. www.dcita.gov.au/crf/paper99/lydia.html (accessed Aug. 20, 2004).

"CAAMA Productions." Central Australian Aboriginal Media Association. http://
www.caama.com/au (accessed Aug. 25, 2004).

"Community Broadcasting in Arnhem Land." *The Media Report,* ABC Radio, July 4,
1996. http://www.abc.net.au/rn/talks/8.30/mediarpt/mstories/mr040796.htm
(accessed Aug. 11, 2004).

"*Cosmic Voyage*: Production Firsts and Facts." National Air and Space Museum.
http://www.nasm.edu/NASMDOCS/PA/CV/cosmicft.html (accessed April 20, 1998;
printouts of page in author's possession).

Cowen, Ron. "Earths Beyond Earth." *Science News Online,* March 1, 1997. http://
www.sciencenews.org/sn_arc97/75th/rc_essay.html (accessed May 3, 1998;
printouts of page in author's possession).

———. "Searching for Other Worlds: A Planetary Odyssey." *Science News Online,*
April 1, 1995. http://www.sciencenews.org/sn_edpik/as_3.html (accessed May 3,
1998; printouts of page in author's possession).

"Declassified Intelligence Satellite Photographs." U.S. Geological Survey, National Mapping Information–EROS Data Center. http://edcwww.cr.usgs.gov/
dclass/dclass.html (accessed Jan. 9, 1996).

"Declassified Intelligence Satellite Photographs." US. Geological Survey, National
Mapping Information-EROS Data Center. http://mac.usgs.gov/mac/isb/pubs/
factsheets/fs09096.html (accessed Aug. 24, 2004).

Deep Dish TV home page. http://www.deepdishtv.org/ (accessed Aug. 24, 2004).

ECI Telecollaborative Art Projects. www.ecafe.com/getty/table.html (accessed
March 4, 2001).

"Ensure Information Dominance." *The Nation's Airforce,* Oct. 2, 1995. http://www
.af.mil/lib/nations/ensure/html (accessed Aug. 1, 1998; printouts of page in
author's possession).

"Ethnic Conflicts in Civil War in Bosnia: Political Manipulation with Term of Genocide." http://www.balkanpeace.org/cib/bos/boss/boss13.shtml (accessed Aug. 7,
2004).

Federation of American Scientists' Image Intelligence home page. http://www.fas
.org/irp/intelwww.html (accessed Dec. 5, 1996; printouts of page in author's
possession).

"First Space Tourist: Dennis Tito's Flight to Station Alpha." http://www.space.com/dennistito/ (accessed Sept. 11, 2001).

"French Documentaries about Alexandria Underwater Excavation." Egyptian State Information Service, July 28, 1997. http://www.us.sis.gov.eg/online/html/011297a.html (accessed July 20, 1998; printouts of page in author's possession).

"General Overview of the Hubble Space Telescope." Space Telescope Science Institute. http://www.stsci.edu/hst/ (accessed April 17, 1998).

Holt, Mike. "Why We Need a Black Press." *Once A Year: African American Press*, May 1996. http://www.tefnet.org/oay/holt.html (accessed July 20, 1998; printouts of page in author's possession).

"Hubble Finds Cloudy, Cold Weather Conditions for Mars-Bound Spacecraft." Space Telescope Science Institute press release, May 20, 1997. http://hubblesite.org/newscenter/newsdesk/archive/releases/1997/15/ (accessed Aug. 24, 2004).

"Hubble Monitors Weather on Neighboring Planets." Space Telescope Science Institute press release, March 21, 1995. http://hubblesite.org/newscenter/newsdesk/archive/releases/1995/16/ (accessed Aug. 11, 2004).

"Hubble Sees Early Building Blocks of Today's Galaxies." Space Telescope Science Institute press release, Sept. 4, 1996. http://hubblesite.org/newscenter/newsdesk/archive/releases/1996/29/ (accessed Aug. 25, 2004).

"Hubble's Deepest View of the Universe Unveils Bewildering Galaxies across Billions of Years." Space Telescope Science Institute press release, Jan. 15, 1996. http://hubblesite.org/newscenter/newsdesk/archive/releases/1996/01/ (accessed Aug. 25, 2004).

"Hubble's Look at Mars Shows Canyon Dust Storm, Cloudy Conditions for Pathfinder Landing." Space Telescope Science Institute press release, July 1, 1997. http://hubblesite.org/newscenter/newsdesk/archive/releases/1997/23/ (accessed Aug. 25, 2004).

"Hubble's Upgrades Show Birth and Death of Stars; Discover Massive Black Hole." Space Telescope Science Institute press release, May 12, 1997. http://oposite.stsci.edu/pubinfo/PR/97/ero/PR.html (accessed Aug. 25, 2004).

Hunter-Gault, Charlayne. "Testament to Bravery." *Online Newshour*. http://www.pbs.org/newshour/bb/entertainment/edmonia_8–5.html (accessed Aug. 9, 2004).

Huntington, Tom. "The Whole World's Watching." *Air & Space Magazine*. April/May 1996. http://www.airspacemag.com/ASM/Mag/Index/1996/AM/twww.html (accessed Aug. 30, 2004).

"Imparja Television home page." http://www.imparja.com.au/ (accessed Aug. 25, 2004).

Israel, Jared, and Nebojsa Malic. "Falsely Accused: Was the Srebrenica Massacre a Hoax?" April 28, 2000. http://www.emperors-clothes.com/analysis/falsely.htm (accessed Aug. 25, 2004).

"Kyoto Protocol on Climate Change Opens for Signature." UN press release, March 1998. http://www.iisd.ca/linkages/climate/climate.html (accessed March 20, 1998; printouts of page in author's possession).

"Maintaining Aboriginality at Imparja." *The Media Report*, ABC Radio, June 27,

1996. http://www.abc.net.au/rn/talks/8.30/mediarpt/mstories/mr270696.htm (accessed Aug. 11, 2004).

"The Making of *Cosmic Voyage*: Roundtable Discussion with Creators Bayley Silleck, Jeffrey Marvin, John Grower, Eric DeJong and Donna Cox." http://www.nasm .edu/NASMDOCS/PA/CV/cosmicmk.html (accessed April 20, 1998; printouts of page in author's possession).

"Mapping the Treasures." *Nova Online*. http://www.pbs.org/wgbh/nova/sunken/ ruinsmap.html (accessed Aug. 13, 2004).

"The Media and Aboriginal Reconciliation." *The Media Report*, ABC Radio, May 29, 1997. http://www.abc.net.au/rn/talks/8.30/mediarpt/mstories/mr970529.htm (accessed Aug. 6, 2004).

Morris, Gary. "Blaxploitation: A Sketch." *Bright Lights Film Journal*. http://www .brightlightsfilm.com/18/18_blax.html (accessed Aug. 9, 2004).

"NASA's Origins Program." http://origins.jpl.nasa.gov/ (accessed Aug. 24, 2004).

O'Regan, Tom. "TV as Cultural Technology: The Work of Eric Michaels." *Continuum: The Australian Journal of Media and Culture* 3, no. 2 (1990). http://wwwmcc .murdoch.edu.au/ReadingRoom/3.2/EMWork.html (accessed Aug. 20, 2004).

Poggioli, Sylvia. "Scouts without Compasses." *Nieman Reports*, originally published fall 1993. Reprinted in special issue vol. 53–54, 2000. http://www.nieman .harvard.edu/reports/99–4_00–1NR/Poggioli_Scouts.html (accessed Aug. 20, 2004).

Sassen, Saskia. "The Global City: Strategic Site/New Frontier." http://people.cornell .edu/pages/sb24/Global_Tensions/papers/sassen.html (accessed Feb. 1, 2002; printouts of page in author's possession).

———. "Saskia Sassen on the Twenty-First-Century City." Interview by *Government Technology*. http://www.interlog.com/~blake/sassen.htm (accessed Aug. 16, 2004).

Schaefer, Kevin. "NASA's Earth Observation System Data Information System." http://www.asis.org/Bulletin/Apr-95/schaefer.html (accessed Aug. 24, 2004).

"Scientific Visualization in the IMAX Film *Cosmic Voyage*." no date. http://zeus.ncsa/ uiuc.edu:8080//Summers/description.html (accessed April 20, 1998; printouts of page in author's possession).

"Scientific Visualization Reaches New Heights in IMAX Film." Grand Challenge Cosmology Consortium press release, Jan. 16, 1996. http://zeus.ncsa.uiuc.edu :8080//Summers/pressrel.html (accessed April 20, 1998).

"Seeking Our 'Cosmic Roots': The Hubble Deep Field Project." Space Telescope Science Institute. http://hubblesite.org/newscenter/newsdesk/archive/releases/ 1996/01/astrofile/ (accessed Aug. 25, 2004).

Sells, David. "Seeking the Truth in Srebrenica." BBC News, June 30, 1999. http://adamjones.freeservers.com/srebren2.htm (accessed Aug. 25, 2004).

"The Sky's Eyes: Remote Sensing in Archaeology." *NOVA Online*. http://www.pbs .org/wgbh/nova/ubar/tools/index.html (accessed Aug. 25, 2004).

Smite, Rasa and Ieva Auzina. Interview of Marko Peljhan. http://make-world.org/ interview_marko.html (accessed Feb. 7, 2002; printout of page in author's possession.

"Statement on Export of Satellite Imagery and Imaging Systems." White House, Office of the President, March 10, 1994. http://www.pub.whitehouse.gov/uri-res/I2R?urn:pdi://oma.eop.gov.us/1994/3/11/3.text.1 (accessed Aug. 9, 1998; printouts of page in author's possession).

"Tiros-1." http://www.earth.nasa.gov/history/tiros/tiros1.html (accessed Aug. 11, 2004).

Treasures of the Sunken City. Transcript. nova *Online.* http://www.pbs.org/wgbh/nova/transcripts/2417treasures.html (accessed Aug. 16, 2004).

United Nations, Office of the Secretary-General. "Report of the Secretary-General on Bosnia and Herzegovina." s/1995/755. Aug. 30, 1995. http://www.un.org/Docs/secu95.htm (accessed Aug. 24, 2004).

"What Is seti?" The seti League, Inc. http://www.setileague.org/general/whatseti.htm (accessed Aug. 25, 2004).

Young, Gale. "Divers Recover Ancient Wonders." cnn *Online,* Oct. 4, 1995. http://cnn.com/world/9510/egypt_treasure/index.html (accessed Aug. 25, 2004).

Index

Numbers in italics indicate illustrations.

Lisa Parks is associate professor of film and media
studies at the University of California, Santa Barbara.

Library of Congress Cataloging-in-Publication Data

Parks, Lisa.

Cultures in orbit : satellites and the televisual / Lisa Parks.

p. cm. — (Console-ing passions)

Includes bibliographical references and index.

ISBN 0-8223-3461-5 (cloth : alk. paper)

ISBN 0-8223-3497-6 (pbk. : alk. paper)

1. Television broadcasting—Social aspects.

2. Direct broadcast satellite television.

I. Title. II. Series.

PN1992.6.P37 2005   302.23′45—dc22   2004025778